Lecture Notes Electrical Engineering

Volume 17

Dirk Henrici

RFID Security and Privacy

Concepts, Protocols, and Architectures

Springer

Dr. Dirk Henrici
University of Kaiserslautern
67653 Kaiserslautern
Germany
henrici@informatik.uni-kl.de

ISBN: 978-3-540-79075-4 e-ISBN: 978-3-540-79076-1

Library of Congress Control Number: 2008924615

Cover design: eStudio Calamar S.L.

Printed on acid-free paper

9 8 7 6 5 4 3 2 1

springer.com

Dedicated to all dear people
accompanying me on the journey of life

Preface

In the beginning of 2003, I found a short article about the privacy implications of RFID technology in a newspaper. It raised my interest, and after reading some early research papers on the topic, I thought: "There must exist better solutions." I concerned myself with the topic in my spare time. After having developed my first solutions, I asked my supervisor, Prof. Dr. Paul Müller, whether I could write a paper about my results. As the topic did not fit into any running project or at least the overall research directions of his group, he could have answered no. But instead, he encouraged me to do it. The paper became a success, and many other papers about new concepts and solutions followed. Now the answer is obvious: There exist better solutions.

I have dealt with the topic over the past years. Now I want to share the basics as well as current research results with the reader. This book is surely not a bedside reading. But with all the presented concepts, it can broaden the mind of the reader concerning security, privacy, and RFID systems. I wish the reader many new insights.

There are many people I would like to thank. First of all, my thanks go to my supervisor, Prof. Dr. Paul Müller. He gave me room for creativity and plenty of rope to work on my own. Thereby, he always trusted in my skills, gave helpful hints, and his door stood open for me at all times. He also suggested the publication of this book. Thanks also go to various persons that supported me in finishing this work. As representatives for all of them, I'd like to mention Prof. Dr. Friedemann Mattern for spending his rare available time on rating my research and encouraging me to go on working, Prof. Dr. Karsten Berns for the valuable discussion about his experiences, and Dr. habil. Bernd Schürmann for answering all my legal and administrative questions. I also want to thank the Springer publishing company for the pleasant cooperation and for publishing this book.

Many thanks go to my present and former colleagues for the pleasant and productive atmosphere as well as the lively discussions on scientific and other topics. I am glad that we had so much fun despite of the high workload. I also want to thank our courteous secretaries who disburden us from much administrative work.

My thanks also go to my friends, Michaela Keßler, Sascha Paulus, Nadine Sumser, and the many others, for their long lasting friendship and their understanding that I was not as available as usual in the recent months.

I am deeply grateful to my parents, Doris and Franz-Josef Henrici, for their enduring support and all the other gifts that I received from them. I thank my sister Carina for distracting me from my work every now and then and for always giving me a hearty welcome each time I visited her.

Very special thanks go to Annika Kohlhaas for her love and patience. She did not only show understanding for all the evenings and weekends that I spent sitting in front of my notebook but also did the proofreading of papers and this book. She is very special, and I am glad that she came into my life.

Kaiserslautern, February 2008 *Dirk Henrici*

Contents

Outline

In the first chapter, the topic of this book is classified into the area of pervasive computing. Further, security and privacy in the scope of RFID technology is motivated and the vision that guides the remainder of this book is introduced.

Chapter 2 begins with an overview of RFID technology. It gives required background information for understanding RFID system components, their interplay, and their relevant characteristics. After a short subchapter regarding security, the core topic of the chapter is addressed: privacy. An introduction to privacy with respect to RFID systems is given with the goal to derive design guidelines. The chapter closes with an introduction to important cryptographic primitives. The focus is laid on one-way hash functions and random number generation since these are the most important primitives within this book.

The successive chapter 3 introduces into the security and privacy issues regarding RFID systems. After motivating the topic using some examples, the variety of threats that the systems is exposed to is identified. Based on these threats, goals for secure and privacy-respecting RFID systems are derived. Reaching these goals has many challenges, which are presented in the next subchapter. Afterwards, security and privacy in RFID systems is considered from an attacker's point of view: Classes of attacker capabilities are introduced and possible attacks on RFID systems presented. The chapter concludes with a presentation and an assessment of the current situation in today's application of RFID.

In chapter 4, concepts for addressing the identified issues are presented. Following the current state-of-the-art in the literature, the presentation focuses on securing the communication between RFID tags and readers and on protecting location privacy despite eavesdropped or unauthorized tag queries. The different concepts are presented along a scheme for classification. As a comprehensive example of an RFID protocol that implements all tasks that are required for secure, privacy respecting RFID systems, the "Hash-based ID variation" protocol is presented and discussed. An evaluation of this protocol is performed based on evaluation criteria that have been proposed before in this chapter.

Chapter 5 addresses a deficiency that has been found in RFID protocols that implement identifier modification based on message exchanges: For these protocols to operate, a central entity is required which limits scalability of the approaches. The problem is addressed using "pseudonymization infrastructures" that enable the building of distributed, inter-organizational systems that have the required security and privacy properties. After motivating the use of such infrastructures in RFID systems, the use of classic concepts for pseudonymization infrastructures in RFID systems is presented. Afterwards, pseudonymization infrastructures based on hash functions as cryptographic primitives are developed.

In chapter 6, the necessity to extend the classic RFID system model that is presumed in current RFID literature and that is also used in chapter 4 is explained. After a discussion of the deficiencies of the classic model, that model is extended step by step: First, readers are regarded as untrusted entities. This leads to a push concept. RFID systems using this new model adhere much better to the practical requirements in inter-organizational systems than ones using the classic model. In a second step, the tag bearer is introduced as separate entity and the concept of a "personal manager" for managing tagged items is presented. The chapter concludes with an evaluation of the complete architecture with the presented protocols and concepts as building blocks.

As the result are RFID systems that fulfill the requirements regarding security and privacy perfectly but that cannot be implemented economically, chapter 7 aims at more practical solutions and presents current research results. At first, a more elegant replacement for the "Hash-based ID variation" protocol is considered. It is called "Triggered hash chain" protocol. Second, a partial solution for securing supply-chains against counterfeited products is presented. The main part of the chapter is the development of the "ID-Zone architecture". It is an RFID system architecture that follows the practical requirements identified in chapter 2 instead of providing privacy in a perfect manner. In return, the architecture can be implemented much more economically. The chapter closes with some research directions.

1

Motivation and Vision

In 1956, the Nobel Prize in physics was awarded to William B. Shockley, John Bardeen, and Walter Brattain for the invention of a functioning transistor. Their achievement, which can be seen as the most important invention of the 20th century [Rio05], set the foundation for a matchless technological development in the history of mankind: The size of transistors became smaller and smaller, the price of transistors lower and lower. This development took place and even today still takes place in an exponential manner so that a current CPU (central processing unit) with a cost of a hundred dollars and a die size of less than two square centimeters has about one billion gates now[1]. The empirical observation of the exponential development [Moo65] is attributed to Gordon Moore, a co-founder of Intel, and therewith the rapid increase in the complexity of integrated circuits is called "Moore's Law".

The rapid progress in integrated circuits had far-reaching consequences: Computational resources became increasingly powerful and cheaply available so that many electronic devices make use of them today. For instance, integrated circuits or even complete embedded systems can be found in cars, in consumer electronics, and in home appliances (e.g. washing machines). The trend that many objects around us are equipped with complex integrated circuits continues and realizes the vision of "ubiquitous computing" which presumes that computational capabilities become ubiquitous. Mark Weiser was one of the people that formed this vision. But besides the ubiquitous availability of computational devices, the vision contains much more:

"The most profound technologies are those that disappear.
They weave themselves into the fabric of everyday life until they are indistinguishable from it."
Mark Weiser, 1991

This citation from [Wei91] focuses on another aspect of computation in the future: Here, not the capabilities stand in the foreground but the unobtrusiveness of the devices. To be unobtrusive, devices need to be "smart" enough to sense context and

[1] source of numbers: Intel Corporation

to anticipate the goals of the users. This is well expressed by the term "ambient intelligence" [Aho01]/[Aho02], which is today often used more or less synonymously to the terms "ubiquitous computing" and the industry-initiated "pervasive computing", respectively. An early publication from Satyanarayanan [Sat01] describes the vision and challenges of pervasive computing and its development history.

The devices are often not stand-alone but are networked. A connection to the Internet, either directly or via a gateway device, enables sharing sensor data of the devices and also enables controlling the devices from hosts in the Internet. This internetworking of devices that are built into everyday items envisions an "Internet of things", a vision that has for instance been promoted by Fleisch and Mattern [FM05].

The goal of the stated visions is always the same: One wants to harvest and share information, process it, and ultimately provide a service. This service could be to give humans assistance in the actions they take. With all the "smart" items being part of everyday life, life shall become easier, safer, and much more comfortable. From an economic point of view, productivity shall increase considerably. Envisioning how the world with such devices will look like in concrete, is theme of movies and lots of publications, but eventually the variety of applications and all the looming possibilities of the technology still remain unimaginable, see also [Mat01].

A very important technology in the context of ubiquitous computing is "Radio-Frequency Identification", abbreviated "RFID". Arbitrary objects, e.g. assets, can be equipped with RFID tags – kinds of labels that can be used to uniquely identify the objects and that can be queried contactlessly and without a line of sight from some distance. Due to the miniaturization of integrated circuits and the decreasing cost, RFID tags could become ubiquitous in the near future and be used in a variety of applications.

In a more abstract view, RFID tags link objects in the "real", physical world to virtual counterparts, e.g. data in databases. This is illustrated in figure 1.1.

Fig. 1.1. RFID bridging real and virtual world

Already this alone provides for a variety of applications, e.g. RFID can act as a replacement for optical barcodes. If tags have additional features (like additional storage or sensors) besides a means for identification, the looming possibilities become even much greater. RFID systems become thus an indispensable part in the vision of the "Internet of things" and can greatly raise comfort and productivity.

But there are also negative sides of RFID technology as well as of the technological progress in general. These sides need to be considered carefully and be dealt with. It is obvious that technological advances and their deployment can and will change our way of life completely like it has happened all over history. The impact upon society can be positive or negative depending on the way the technology is used. Some examples for this will be given in the following.

Humans nowadays depend on the functioning of their technology. For instance, if a large part of the power grid or the Internet broke down due to natural disasters, war, or terrorist activity, the economic impact would be enormous and the life of people would be endangered. One is used to the functioning of the technology and a sufficient preparation for a longer malfunctioning is very difficult to achieve. Whoever can control or harm the important infrastructures has enormous power. Thus, as people rely on the proper functioning of the technology, characteristics like reliability, safety, and security of the built systems are of very high importance.

As the envisioned systems collect, store, and process a vast amount of data, including personal information, privacy becomes important as well. We need to consider how much privacy we want to drop in exchange for the possibilities the systems offer. Here, a proper balance needs to be found and the societal impact of any changes needs to be considered carefully.

Besides the impact on society, there are of course other consequences like the impact on the environment. Raw materials are needed for manufacturing devices and the materials should be properly disposed after use. Furthermore, a lot of energy is needed, not only for manufacturing but also for operation within the product life. Imagine how much power is needed to operate the Internet world-wide: Routers, switches, servers, etc. all over the world require an enormous amount of power for their operation.

Consequently, beyond the technological matters there are a lot of other aspects, e.g. ethical or societal ones, to consider. The RFID technology is a technology that has the potential to change our way of life completely. This book deals mainly with the security and privacy aspects regarding RFID. The goal is to find dependable technical solutions for the issues caused by the use of that technology.

In recent years, RFID became a "hype" technology. Many companies have already started to use the technology, first in pilots, later in productive environments. The rapid increase of RFID sales leaves no doubt that RFID will gain much more importance in the years to come. Due to the huge number of possible applications and the decreasing costs for tags, the technology will inevitably become ubiquitous in the near future.

But besides the positive possibilities and the high expectations there are also fears that the technology could be abused. Newspaper headlines speak a clear language: Titles like "Are you wearing track shoes?", "Cradle-to-Grave Surveillance", or "RFID tags: Big Brother in small packages" are all but unusual. Not all of the articles are objective, but the message is clear. In sum there are two fears: One is that RFID could be used for creating very detailed customer profiles leading to vast amounts of information that are used for marketing purposes. There are many scenarios imaginable in which such information could be abused. The second fear is that RFID could become an instrument for keeping people under surveillance. It could be the technological basis of a surveillance society that would outplay the scenario known from George Orwell's famous book "Nineteen Eighty-Four" [Orw49] considerably.

A simple example of an unwanted privacy violation is the following scenario: If a person carries clothes, books, pharmaceuticals, banknotes, and other items all of which are equipped with unsecured RFID tags, anybody could read out the data from the tags unnoticed by the person just by passing by with a reader near enough. The brand of the clothes and other personal items could give information about the financial situation of the person; the information read about the books could reveal the person's interests or even problems (imagine a book title like "Alcoholism at work"); the data about the pharmaceuticals could give information about the health status (diseases like infection with HIV); the amount of money the person carries would be interesting for a pickpocket; the content of a woman's handbag would be revealed.

The problem with the RFID technology is that it has pros and cons. On the one hand, it has many applications making life more comfortable and companies more productive. But on the other hand, it can have negative implications regarding privacy. Another problem with most current RFID systems is that security is not addressed properly, see e.g. [HJSW06].

The ideal state would be that the power of RFID technology could be used without causing problems regarding security and privacy. This state is the vision guiding this book. The theory is that security and privacy can be ensured using technical safeguards if the whole RFID system is designed properly. A security and privacy by design approach is demanded because otherwise the goals cannot be fulfilled, see [HM05] and [Lan01].

This vision of a world in which privacy persists and security is ensured but the full potential of the technology is nevertheless tapped shows the research directions. The goal is to find technical ways to provide security and privacy in RFID systems within the various given constraints: technically and economically but also ethically and socially. The solutions that shall be found need not only provide security and privacy but also need to be reliable, scalable, flexible, inter-organizational, and lasting. This is an immense challenge.

2

Fundamentals

This chapter introduces topics that are essential for exploring the research area of security and privacy in RFID systems. At first, an overview of RFID technology and the involved system components is given. In the subsequent section, some relevant aspects regarding security are presented. In a third section, privacy is discussed with the goal of deriving design guidelines that can be used for RFID systems. Afterwards, cryptographic primitives that are relevant for RFID systems are stated, and relevant characteristics are discussed.

2.1 Radio-Frequency Identification

Radio-Frequency IDentification, abbreviated "RFID", basically provides a means to identify objects having RFID tags attached. Fundamentally, RFID tags provide the same functionality as barcodes but usually have a globally unique identifier. Using RFID, the identification is performed electromagnetically. Thus, there is – in contrast to barcodes – no line-of-sight necessary, and the identification can also be performed contactlessly. RFID also has the advantage that bulk reading is possible and that it is not susceptible to dust, dirt, or vibration like barcodes. Because of these characteristics, RFID is envisioned to be a convenient replacement for optical barcodes in the future. Unfortunately, RFID also introduces problems: It is simple to disrupt service, and due to the convenient reading, problems respecting data security and privacy arise.

Today, the term "RFID" is also used in a broader sense. If RFID tags have extended functionality like data storage, computational capabilities, etc., they can have smartcard-like functionality; if RFID tags have sensors, e.g. for temperature measurement, they can also be used for telemetry applications.

Auto-id systems provide automatic identification of objects. RFID systems are a subset of such auto-id systems. This is depicted in figure 2.1. Depending on the provided functionality, RFID systems can be classified somewhere in-between barcode systems and smartcard-based systems.

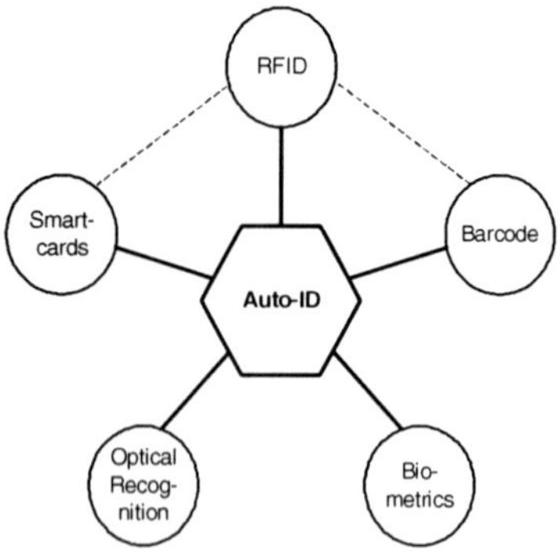

Fig. 2.1. Classification of RFID as auto-id system

RFID systems have more components than the already mentioned RFID tags, just like barcode systems consist of more than just the printed barcodes. A distinction of the following three components is common (e.g. [PLHCETR06], [OSK03], [CLL05]):

- RFID tags,
- RFID readers, and
- backend systems (i.e. middleware and applications).

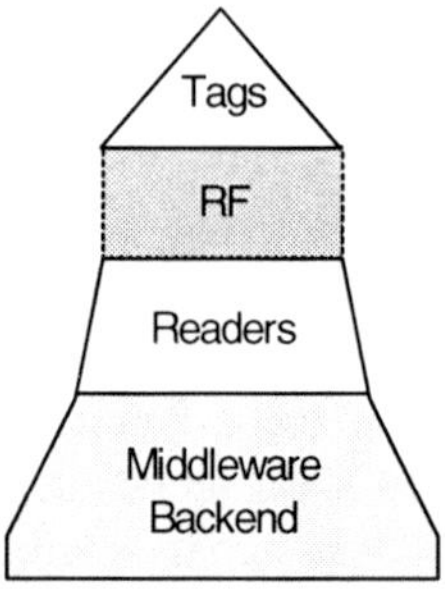

Fig. 2.2. RFID system components

These components are depicted in figure 2.2. Tags and readers communicate over a shared, insecure RF-channel.

In some publications (e.g. [Fel03], [Fin03]), the backend as data processing infrastructure is neglected, although it is of major importance for effective RFID application and although it is of high potential for future innovations. Other publications at least mention the backend in form of a database that is accessed by the reader (e.g. [SWE02], [WSRE03]).

As a good understanding of the RFID technology is essential for hte remainder of this book, in the following subsections, the relevant aspects of the technology are discussed in detail. After a very short historical survey, the components of RFID systems are discussed one after another. The section concludes with some considerations and background concerning the complete systems. For more detailed information about RFID see the book [Fin03] (a German edition exists, too).

2.1.1 RFID History

RFID is not a new technology. It was first used in military. In World War II, it was used for an application called "Identify Friend or Foe" (IFF). The idea was to gather whether an aircraft was friendly or an enemy by the detection of a radio signal which was emitted by the peer upon interrogation. Afterwards, identification technology was developed for tracking military equipment and personnel.

In the 70th, commercial forerunners for retail security systems appeared, and standards for cattle identification were created. Starting in the 80th, the technology was widely used for cattle identification in Europe and toll collect systems were developed. In the 90th, the technology was already present in many kinds of applications, e.g. for toll collection or access control systems. At the beginning of the current decade (20xx), RFID became a "hype" technology. RFID tags were seen as the successor of optical barcodes and companies tried to increase productivity with the technology. As the technology was not mature enough for wide-scale application, a time of disillusion followed. Currently, RFID technology establishes fast, but the euphoria seems to be over.

2.1.2 RFID Tags

RFID tags are also called *transponders*. Both terms are in use and utilized synonymously. The term *tag* places emphasis on the use as a label and the application for item identification. In contrast, the term *transponder*, a connection of "transmission" and "respond", puts stress on the communication process, i.e. answering the request of a reader.

Composition and mounting forms

An RFID tag is composed of the following three components:

- antenna,
- microchip, and
- encapsulation/packaging.

Active RFID tags additionally have a power supply.

A lot of different forms of mounting exist for RFID tags. So-called *adhesive labels* that can be treated like stickers are most common. Examples for mounting forms are the mentioned adhesive labels, card transponders, glass cylinders, plastic packaged transponders, and transponders in robust industrial packaging.

Materials

Nowadays, the usual semiconductor materials like silicon compounds, copper, and aluminum are used for antenna and microchip. Some companies perform research in order to enable creation of RFID tags that are solely based on polymers, i.e. organic materials. One wants to become able to simply print out RFID tags and thus to save costs because assembly of different parts would be no longer required. Organic tags that are biodegradable would also solve the problem of environmentally friendly disposal of old tags.

Power supply

There are two completely different kinds of tags regarding power supply: Passive ones and active ones. Passive tags do not have an own source of power (i.e. battery) and thus must harvest the needed energy from the electromagnetic field of the reader. In contrast to that, active tags have an own power source. There also exist hybrid forms, i.e. semi-active tags. This means that the tags have an own power supply for the microchip but communicate using the power of the field of the reader.

Passive tags are much cheaper than active ones and thus have the greatest market share. Their disadvantages are a rather limited read range and limited functionality; for instance, there is no power for continuously monitoring products using sensors in the tag. Advantageous are their low costs, their small size, low weight, and an economic lifetime that is not restricted by battery life, which allows long-lasting service. In contrast, active tags can communicate over longer distances and can have more functionality but have a limited life, a higher weight, and a higher price. Thus, if the specific characteristics of passive tags are suited for the particular application, they are the best choice. In this book, only passive tags are of concern because of their future ubiquity and their challenging resource constraints.

Functionality

The memory of RFID tags has a size of a single bit (anti-theft devices) up to several kilobytes. Tags with 96 bit = 12 byte for storing an EPC ("electronic product code", see section 2.1.6) are very common.

A distinguishing feature are the employed *memory technologies* that can also be used in combination:

- non-volatile storage: read only (fixed after manufacturing),
- non-volatile storage: WORM (write once, read many),
- non-volatile storage: read/write, and
- volatile storage (for performing calculations after power-up).

WORM tags are written when they are applied. For instance, they get an application specific identifier.

Besides functionality for reading and storing data, RFID tags can have *additional logic* for performing calculations. For the field of security, computational functionality can range from password check for certain operations over implementation of hash functions (see section 2.4.3 below) up to implementation of ciphering algorithms.

There are also tags that are equipped with *sensors*. Ones with buttons (e.g. for activation for use as keys) and with temperature sensors (e.g. for monitoring cold chains) are common. Therewith, applications range into the field of telemetry.

Passive tags with *displays* are currently in development. This technology was presented by EPSON at Auto-ID Expo Tokyo in 2005, see [Web05], and is regarded to have a very high potential ([CZ06]). It is not a problem to drive standard displays with active tags, but when using passive tags there is no permanent power supply. The idea is to use "ePaper" as display technology because it maintains its content (e.g. product price) without power.

Costs

The costs of RFID tags are often the decisive element in RFID systems because the tags are needed in large quantities. The price for a low-cost tag, i.e. a simple tag with few memory and almost no computational capabilities, lies approximately between ten and fifty cents depending on type and quantity but is expected to drop well below five cents in the near future. According to [KC04], manufacturers predict that they can bring down the price down to about one cent for the cheapest tags within five years.

Standards

Most of the existing standards regard the communication between tags and readers and ensure interoperability between devices of different vendors. These standards will be listed in subsection 2.1.4. Standardization of tags (application level) is not very important for system operation because flexible software can easily deal with different kinds of tags. But on the other hand, it would be nice if there were not too many different standards because

- higher quantities of tags of the same kind lead to lower costs per tag,
- no additional pieces of software for dealing with different kinds of tags is required, and
- fewer different kinds of tags lead to better privacy protection and higher transparency for customers (e.g. no different levels of protection concerning data security and privacy).

There are only few international standards concerning the configuration of tags in certain applications. Standards of the ISO (International Organization for Standardization) are: Radio frequency identification of animals (ISO 11784/11785:1996), Radiofrequency identification of animals – Advanced transponders (ISO 14223-1:2003), and Gas cylinders – Identification and marking using radio frequency identification technology (ISO 21007:2005). These standards do not only specify the application level tag configuration but the whole stack including the communication interface characteristics.

Besides those standards, there is a widely adapted specification of EPCglobal Inc. that describes the functionality that tags need to provide for delivering the EPC (Electronic Product Code). This specification defines the complete stack from the air interface, i.e. communication between tags and readers, up to the configuration of tags (states, high-level protocols, etc.) and the EPC itself. More information regarding this specification can be found below section 2.1.4.

Research topics

Some current research topics regarding RFID tags are the following:

- materials (for decreasing price and easier disposal),
- optimization of the manufacturing process,
- power supply (efficient use of energy, reliability in case of fluctuations and power outages),
- efficient use of tag resources,
- memory technologies (in particular non-volatile storage, e.g. number of possible write cycles), and
- efficient implementation of algorithms in hardware (e.g. cryptographically ones).

2.1.3 RFID Readers

RFID readers send and receive data to and from tags. Because of that, they are often also called "transceivers", a shortened concatenation of the words "transmitter" and "receiver". RFID readers are the connecting element between the RFID tags and the middleware or backend systems. Thus, they consist of an antenna along with the required electronics for communication, a microprocessor for controlling the device, and an interface for forwarding the data to the processing backend system. The power supply of the reader is also used for powering passive tags via the electromagnetic field created by the reader.

There are two different categories of readers:

- stationary readers and
- mobile readers.

Stationary readers have a fixed location and a permanent network connection can be presumed. In contrast, mobile readers can be moved around and application scenarios may exist in which no network connection is available. For instance, stationary readers could be located at goods receiving, whereas mobile ones could be used for querying prices of goods in a supermarket or for machine maintenance. Of course, different models of readers exist: For example, "gates" can be used at doors and or "tunnel readers" at belt conveyors.

2.1.4 Communication Between Tags and Readers

One of the most important characteristics of RFID systems in contrast to barcode systems is that the communication with the reader does not need a line-of-sight. There is also no wiring required. Instead, the communication between RFID tags and RFID readers is performed by electromagnetic means. This has advantages but also has inherent problems.

Reference model

Communication in networks is often separated into a number of orthogonal layers for coping with complexity and for better filing of technologies. When talking about the communication between computers, often the ISO/OSI reference model [ISO94] is used. The communication between tags and readers can be discussed on the basis of a simplified model in which the lowest two layers correspond to the ones in the ISO/OSI reference model and a third layer corresponds to all upper layers in the ISO/OSI reference model, see figure 2.3. A similar layering has been presented in [AO05a].

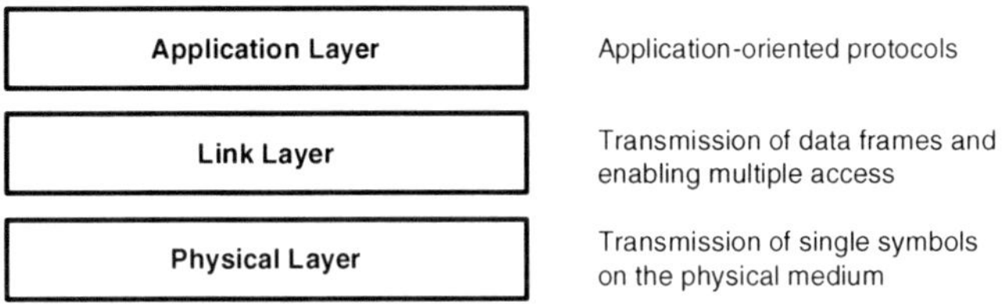

Fig. 2.3. Layering of communication between tags and readers

The lowest layer, which is called *physical layer*, defines the transmission over the physical medium, i.e. used frequencies, modulation techniques, signal forming, etc. Details about physical layer issues can be found in [Sch01]. Above that layer, the *link layer* is placed, in which the transmission of data frames, i.e. sequences of bits belonging together, occurs. At this layer the algorithms for multiple access to the shared medium can be found. Protocol messages, e.g. for reading and writing the memory of tags or authentication algorithms are situated in the upper *application layer*.

The lowest two layers need to be standardized to a large extent to ensure interoperability between tags and readers of different vendors. These needed standards already exist, see below in subsection 2.1.4. But on the application layer, there is freedom for own developments, i.e. user specific tags, without breaking interoperability.

Frequency ranges

The following frequencies are widely used for communication between tags and readers:

- LF / low frequency / 125-134 kHz,
- HF / high frequency / 13.56 MHz,
- UHF / ultra high frequency / 868 MHz and 915 MHz, and
- Microwave / 2.54 GHz and 5.8 GHz.

Apart from these frequencies, there are some other frequencies in the ISM band ("industrial scientific medical" band) that are free for use, e.g. 433 MHz. But these frequencies are not widely used for RFID applications.

Different frequency ranges have different physical characteristics that, for instance, affect the needed size of antennas or the read range. Besides that, materials like metals and fluids have different influence depending on the frequency range. For example, HF tags are significantly affected by conductive materials, whereas at higher frequencies (UHF, microwave), there emerge problems with absorption, reflection, and refraction. Tags that communicate using high frequencies are usually active, thus having an autonomous power supply. In sum, the ideal frequency range to be used depends on the application, e.g. on the environment the communication takes place in.

Electromagnetic fields

One can distinguish between two kinds of fields used for data transmission and powering the tags:

- near fields (magnetic and electric fields, i.e. inductive/capacitive coupling) and
- far fields (electromagnetic waves).

The maximal field strength allowed depends on national regulations. It is limited for electromagnetic compatibility (EMC), i.e. for avoiding disturbing other systems and to prevent harming nature and environment. Passive tags operating in the near field use load modulation to send data to a reader, ones operating in the far field use backscattering.

Multiple access and anticollision algorithms

Techniques are required that ensure that several devices, i.e. several tags and readers, can communicate using the same shared medium without disturbing each other. Tag anticollision is especially important for bulk reading of tags; reader anticollision is comparatively new [ES02] and becomes important due to the occurring increase in reader density.

As the used frequency is fixed and thus FDMA (frequency division multiple access) not a solution, one can perform different communication transactions at different times (TDMA, i.e. time division multiple access). Access to the communication medium can be controlled using two classes of anticollision algorithms:

- probabilistic anticollision algorithms and
- deterministic anticollision algorithms.

Probabilistic algorithms use the principle of the ALOHA-protocol that is the ancestor of the CSMA/CD protocol (collision sensing, multiple access with collision detection) that is still used in the Ethernet. The ALOHA-protocol works as follows: A station, i.e. a device, may start to send if the communication medium is free. If a collision occurs because several stations have started to send data, the medium is released by all stations. Each station waits for a random time and may then try to send again if the medium is free. Probabilistic algorithms are efficient as long as the probability for collisions does not become too high. This means that not too many tags and readers may be located within the transmission range to avoid a high increase of collisions and therewith a rapidly decreasing read rate (number of tags read per time unit).

The second class of anticollision algorithms is the deterministic ones. For the communication between RFID tags and readers often the *binary tree walking* algorithm is employed: The reader controls the communication process. Each tag has a unique address, i.e. a unique identifier. With these prerequisites, the binary tree walking takes place as follows: The reader requests all tags whose address begins with a 0-bit. If multiple tags answer so that a collision occurs, the reader gets more specific and asks all tags to reply whose address begins with "00". The query is made more specific until only a single tag answers. Afterwards, the remaining branches of the binary tree are traversed using the same algorithm. At the end, the reader knows all the addresses of the tags around and can directly address single tags using their respective address.

Read range

The attainable read range depends on many aspects in practice. Some of these aspects are

- employed frequency,
- kind of power supply (active vs. passive tags),
- field strength of the reader or the active tag (regulations limit the permitted field strength),
- antennas (size and form) of the reader and the tags,
- alignment of antennas of tags and readers to each other,
- sensitivity and tuning of electronics in tags and reader (e.g. suppression and filtering of noise and interfering signals),
- energy demand of the integrated circuit (microchip) of passive tags, and
- environment (e.g. ambient materials like metals, fluids, or other tags can lead to detuning).

Read range and reliability (see next paragraph "Reliability") are at odds because the reliability of reading decreases with increasing distance between tags and readers. This is due to the weakening of the electromagnetic fields and electromagnetic waves, respectively.

The physical limit of the range using near field coupling can be approximated to $\frac{\lambda}{2\pi} = \frac{c}{2\pi f}$, where λ is the wave length, c is the speed of light, and f the operating frequency, see [Wan06] and [Wei03]. An additional limitation of the read range using near field coupling occurs due to the decrease of the field strength with the factor $\frac{1}{r^3}$, where r is the distance between the antennas of tag and reader. The effect is that the available power for tag operation decreases rapidly with increasing distance and that the signal to noise ratio (SNR) becomes worse.

The field strength of the far field only decreases with a factor of $\frac{1}{r^2}$. Powering tags is therewith possible at greater distances. As backscattering is used for transmitting data from tags to the reader, the received energy at the reader drops with the factor $\frac{1}{r^2} \cdot \frac{1}{r^2} = \frac{1}{r^4}$ so that a very sensitive receiver is required [Wan06].

Low-frequency and high-frequency tags have a read range up to 1.5 meters (see e.g. [FM05]) but partly only a few centimeters. UHF tags have a range up to about 5 meters. These values base on currently allowed field strengths.

Eavesdropping of communication is possible over much farther distances because the rather short read range of common readers results by the fact that the power for operation of passive tags needs to be propagated using the electromagnetic field of the reader. For UHF tags, the forward channel can theoretically be monitored from a distance of 1 km and the backward channel from up to 100 meters [SWE03].

Reliability

Reliability concerning tag read-out, i.e. the probability that a tag can be accessed (read/write) successfully, depends on the same aspects that have been listed regarding the read range in the previous subsection. With increasing distance between tags and readers, the probability of a correct exchange of protocol messages decreases.

Due to the many aspects and the different application scenarios, meaningful quantitative values for the reliability cannot be given. This is reflected by the fact that numbers given in publications or product sheets vary widely. For RFID systems that shall operate in a reliable and productivity improving manner, the middleware that aggregates and filters the data should account for missing data caused be transmission errors and consider the case that a tag is there but could not be read.

Speed of reading

The achievable speed of reading, i.e. the number of read tags per second, depends on:

- available data rate (depending on physical layer, e.g. on frequency),
- anticollision algorithm and number of tags within communication range,
- error correction strategy and wanted reliability, respectively,
- amount of data to be transmitted,
- number of messages to be exchanged, and
- required time for performing calculations (e.g. for authentication algorithms).

For low-frequency and high-frequency systems (ISO 15693 [ISO01b] and 14223 [ISO03]), the data rate is approximately 5 kbit/s. Newer high frequency systems according to ISO 18000-3 [ISO04] have a data rate exceeding 100 kbit/s. Ultra high frequency systems according to ISO 18000-6 [ISO04] reach about 50 kbit/s. Depending on the used anticollision algorithm and the amount of data to be transmitted, about 10-30 tags per second can be read using LF and HF systems and 100-500 tags per second using UHF systems [FM05].

It makes sense to always keep the number of tags that need to be read as low as possible. This could be done be introducing a hierarchical structure (pallet level, item level, etc.) and to always read only that level that is required in the current application.

Standards

There exists a variety of standards for ensuring interoperability of devices by specifying the physical layer and the link layer shown in the reference model. Except for the limitation that the standards are differently suited for certain applications, e.g. due to the employed frequency and the resulting characteristics, the standards presented in the following are independent of the field of application.

ISO standards focusing on the communication between tags and readers:

- ISO/IEC 14443:2000/2001: Identification cards – Contactless integrated circuit(s) cards – Proximity cards [ISO01a]
  - Part 1: Physical characteristics
  - Part 2: Radio frequency power and signal interface
  - Part 3: Initialization and anticollision

  Communication takes place at 13.56 MHz in this standard.
- ISO/IEC 15693:2000/2001: Identification cards – Contactless integrated circuit(s) cards – Vicinity cards [ISO01b]
  Communication takes place at 13.56 MHz in this standard.
- ISO/IEC 18000:2004: Information technology – Radio frequency identification for item management [ISO04]
  - Part 1: Reference architecture and definition of parameters to be standardized
  - Part 2: Parameters for air interface communications below 135 kHz
  - Part 3: Parameters for air interface communications at 13,56 MHz
  - Part 4: Parameters for air interface communications at 2,45 GHz
  - Part 5: [not listed]
  - Part 6: Parameters for air interface communications at 860 MHz to 960 MHz
  - Part 7: Parameters for active air interface communications at 433 MHz

These ISO standards are widely respected. Even EPCglobal Inc. with its huge support by industry has changed their initial specification to comply with them which led to the "Class 1 Generation 2 UHF Air Interface Protocol Standard Version 1.0.9" [EPC05b], known as the "Gen II" standard.

There are also other standards like Zigbee (IEEE 802.15.4a, [IEEE03]), which operates at 2.54 GHz, that could also be used for RFID purposes. For item management, also the ISO standards 15961, 15962, and 15963 headed "Information technology – Radio frequency identification (RFID) for item management" exist. They define an application specific data protocol and thus operate at a higher layer.

Within this book, it is not relevant which standard is used on the physical layer. The considerations are not bound to any specific characteristic of that layer. Most aspects of the link layer (e.g. framing) are also not relevant for considerations in this book, but there are some relevant ones like addressing because they can affect security and privacy.

Research topics

Some current research topics regarding RFID communication are the following:

- reliable communication in different environments,
- securing the communication between tags and readers,
- efficient and reliable communication protocols,
- optimization of characteristics like speed-of-reading, and
- ensuring location privacy.

Tags are resource scarce environments due to cost constraints. One has to cope with all the resulting limitations. The issues here are subject of this book.

2.1.5 RFID Backend Systems and Middleware

Readers are used for querying tags and reading and writing tag data. All the read data needs to be processed, and the data to be written needs to be available, so that an additional system component is required to form a complete RFID system: the backend. It can be separated into two parts: into middleware and applications. Both are running on computers within the network.

Middleware can be used to aggregate and filter data and to provide an open and neutral interface towards the applications. It can decouple applications and specific tag and reader characteristics (like special protocols, vendor specific implementations) so that an upgrade towards new tag and reader technology can be performed comparatively easily since no adaptation of applications is required. Using middleware, applications can easily cope with "old" tag and reader technology. Using middleware, it is not required that an application is able to handle all kinds of old tags individually (exchanging all old tags by new ones will not be feasible any more when tags are ubiquitous). Moreover, it makes sense to move as much processing from tags into middleware as possible because computation and memory is much cheaper there.

Middleware for RFID can also handle many other kinds of auto-id technology. For instance, the functionality of barcodes is a subset of the functionality that is provided by RFID tags so that a barcode can be regarded as a read-only RFID tag. In contrast, current barcode middleware that is used for processing the read barcodes is not able to handle the more complex requirements of RFID technology.

Middleware should be responsible for:

- data provision for other IT-systems within the organization,
- data exchange to and from other organizations,
- coping with read errors and tags that could not be read, respectively,
- access control for accessing data associated to tags,
- decoupling between readers and applications,
- decoupling between tag layout and applications, and
- gaining flexibility by modularization and decoupling of system components.

This list is not complete. In sum, only functionality that is specific for the application should be handled from the application. Everything also should be handled by middleware to avoid redundant implementation of logic.

Compared to tags and readers, the relevance of backend systems and especially middleware is often undervalued when talking about auto-id systems. The result is that the looming possibilities that lie within the technologies, e.g. optical barcodes, are today by far not exploited.

Standards

There are virtually no standards for middleware. So there is still a gap that leaves a lot of potential for research that leads to specification of RFID data processing. Even for the processing of Electronic Product Code data, only a few aspects are specified. Instead, everything is done in enterprise resource planning systems (ERP; e.g. from SAP) in an application specific and proprietary manner.

For instance, only few aspects are specified by EPCglobal Inc. yet:

- Reader Protocol Standard [EPC06b]
  Original description: "Reader Protocol is an interface standard that specifies the interactions between a device capable of reading/writing tags and application software."
- Application Level Event (ALE) Specification Version 1.0 [EPC05a]
  Original description: "This [...] standard specifies an interface through which clients may obtain filtered, consolidated Electronic Product Code (EPC) data from a variety of sources."
- Object Naming Service (ONS) Specification Version 1.0 [EPC05d]
  Original description: "This document specifies how the Domain Name System is used to locate authoritative metadata and services associated with [...] a given Electronic Product Code (EPC)."
  A short technical description as well as an argumentation why data should be stored in the backend can be found in [SWE02].
- EPCglobal Certificate Profile [EPC06a]
  Original description: "This document defines an X.509 certificate profile for use in the EPCglobal network."

2.1.6 RFID Overall System

Tags, readers, middleware, and applications together form a whole RFID system. In the following, some aspects that are relevant to RFID systems as a whole are discussed in short.

Data management

Arbitrary data can be associated to each object tagged with an RFID tag. This could for instance be the date of manufacturing, minimum durability, batch number, etc. There are two possibilities where this data can be stored:

- directly on the tag or
- in a database in the backend.

If the data is stored in a database, a tag only needs to have a unique identifier that can be used as key within the database so that data can be linked to the tag. Such a database does not need to be a central one but can be partitioned amongst multiple organizations.

Advantages of storing data off-tag in a database:

- cost savings: storage space on tags is much more expensive;
- data can be easily changed without the tag being in range of a reader;
- data can be queried without the presence of the tags storing it;
- decoupling of subsystems results in simpler migration paths towards future applications;
- interoperability can be guaranteed more easily since the backend has the capabilities to deliver data in the needed format independently from physical storage;
- more flexible and extensible access control is possible as there is no lack of resources;
- easily upgradeable (augmenting additional data, new security primitives, etc.);
- data security can be ensured more easily (better access control, no data needs to be transmitted over the insecure air interface between tags and readers).

Advantages of storing data on the tag:

- mobile applications are implementable easily;
- simpler system architecture for simple applications;
- ability to store data immutably and in a distributed manner.

Due to the stated advantages of data storage in a database, this possibility should be preferred within most application scenarios, although it seems to be the more complex and more far-fetched approach at first sight.

The limited resources in low-cost tags prohibit the storage of large amounts of data and do not give the possibility to implement flexible security mechanisms that can be used in the backend without constraints. Thus from a security and privacy point of view, data storage in a database is the presumed method in most publications regarding this topic. The only exception is if completely distributed data storage is wanted or data should be absolutely immutable: For instance, biometrical data is stored in passports instead of a central database so that nobody – not even the operators of the database – can query or steal the data[1]. As already stated, storing data within the backend infrastructure also has the advantage that data can be accessed and altered without the tag being present. Backup of data and data recovery of defective tags is only viable when the data is stored in the backend.

Fields of application

There are very many applications for RFID technology. Many of them cannot even be imagined today but will emerge when RFID tags and readers become ubiquitous. Figure 2.4 tries to categorize the major fields of application and shows the important features of the technology that these fields of application are based on.

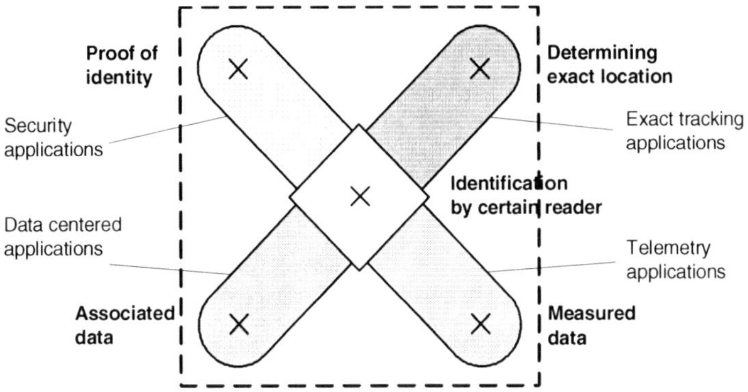

Fig. 2.4. Core areas of major fields of application

All applications require a tag to identify itself. This is depicted in the center of the figure. This identification takes place by a query performed by a reader. As the location of many readers is fixed, the rough location of the tag is also identified. The identification by a certain reader is the "core" functionality of tags and is already sufficient for some kinds of applications like the prevention of shop lifting by affixing tags that are read by readers at the shop doors or for time measurement in sports by identifying a runner by an RFID tag read at the finish line.

Another area of application is security and access control. Besides the identification, it is relevant to prove that the proclaimed identity is correct. Example applications are identity cards and passports, payment (toll collection, micro payment, next generation credit cards), access control to restricted areas, vehicle immobilization systems, anti-counterfeiting (e.g. for medicine or bank notes), and theft prediction/detection.

The currently largest area of applications is the one in which additional data is associated to the identified tag. This is relevant to logistics, i.e. supply chain applications with the aim of tracking and tracing of assets, reducing out of stock, speeding

[1] a change of this design decision towards storing data twice is under discussion in Germany

up delivery, helping in produce to demand, faster recalls, etc. Another goal of applications relying on data associated to a tag is to give assistance to people and automate processes in different areas of life – at home, at work, as well as in public. One vision is the "intelligent home" with "smart" appliances like a microwave oven that automatically detects how to cook a convenience food, a washing machine that checks whether its content may be washed all at once, or a refrigerator that checks the freshness of goods. RFID could also help to organize items at home, e.g. the personal CD collection or the personal library, or be used in games, e.g. for detecting cheating and giving assistance in card games. Especially for elder people, applications from "smart" healthcare (e.g. a "smart" medicine cabinet) up to assisted living become relevant. Documents in the office could be equipped with tags, so that, for instance, bills have directly stored the bank connection and the data required for book keeping so that such information no longer needs to be entered manually. Further, the working place, vehicles, and many more objects could be personalized with tags to which a user profile is assigned. Department stores could work checkout-lessly if all assets had tags affixed. Other applications among many others are recycling (objects with tags "know" which materials they are composed of), animal identification (pets, livestock, etc.), and toll collection.

If tags are also equipped with sensors, the possible applications range into the field of telemetry. As already stated in section 2.1.2 about tag functionality, tags can have a variety of sensors. For instance, an application could be to check whether cooling chains were not interrupted and thus to ensure the freshness of products or the effectiveness of medicine.

A fourth field of application is one in which an object with a tag is identified and its location is determined with particular accurateness, i.e. more precisely than could be gathered by the fact that a tag was read by a certain reader with a specific location. Applications would be, for example, tracking and finding assets in a factory or precisely locating people in a building.

Of course, there are applications that cannot be clearly assigned to one of the stated core application areas because they use tags with a several optional features.

A numbering scheme: Electronic Product Code

Numbering schemes are required for structuring the numbering of tags. The most common scheme is the "Electronic Product Code" (EPC). It is a superset of a variety of numbering schemes. The EPC is intended to replace many of the current schemes like the "Universal Product Code" UPC) of the "Uniform Code Council" (UCC) or the family of "European Article Number" (EAN) codes that are widely used today in conjunction with optical barcodes.

The EPC is specified in the "EPC Tag Data Standard Version 1.1 rev 1.27" [EPC05c]. According to the specification, an EPC has 64 or 96 bits. The first part of each EPC is a header that defines what fields the rest of the EPC is composed of, i.e. which scheme is used. For instance, a "Serialized Global Trade Item Number" (SGTIN) consists of the following parts: filter value, partition, company prefix, item reference, serial number. Such parts are found in many numbering schemes in a similar manner. "Filter value" is used to limit the tags that need to be read, "partition" defines how many bits are spent for each of the remaining code parts, "company prefix" defines the organization which a tagged product belongs to, "item reference" defines the type of the tagged product, and "serial number" makes products of the same type distinguishable. There are also schemes that do not distinguish between item type and serial number or ones that do not have fields for defining partitioning.

The big difference to the wide-spread numbering schemes that are in use today is that the higher number of available bits gives space for a serial number that enables to distinguish objects uniquely.

2.1.7 RFID Summary

This section introduced the RFID technology and gave an overview on relevant aspects of that technology. Different system components have been presented and explained. Some topics that are important for the following chapters have been highlighted. For instance, it has been shown that keeping data in the backend is generally preferable.

The solutions that are designed and discussed within this book are independent from the link layer technology and the layer below if not explicitly stated otherwise. In particular, the physical characteristics of the air interface, e.g. used frequency, are not relevant regarding the security and privacy considerations.

2.2 Security

In the field of information security, the term "security" is used in different contexts and in different senses. One the one hand, it can be applied to systems or services; on the other hand, it can be used in connection with data or information. Both are related to each other, because data security can only be ensured by secure systems. Therefore, the term "information systems security" is widespread as general expression. The term "security" can denote the "secure" condition of a system, services, or data but sometimes also the safeguards to achieve that condition.

The security of a system expresses the ability of a system to behave in the way it is intended to – even in the presence of conditions or even deliberate hostile efforts, i.e. "threats", that intend to make it misbehave. This definition is related to the one given in the "Department of Defense Dictionary of Military and Associated Terms" [DOD94]. In short, one can summarize the term "security" as "a condition in which an entity does not suffer harm from threatening events" [Cla01]. Within this book, secure RFID systems shall be discussed and built, i.e. RFID systems that operate as intended even when threatening events occur, e.g. attackers try to do harm to these systems.

2.2.1 Properties of Secure Systems

Information security is usually benchmarked using the categories “confidentiality”, “integrity”, and “availability” (see [ISO05]). When appropriate, additionally “privacy” (see next section) or more application specific aspects like “non-repudiation” (e.g. for online transactions) are used. In the following, only the usual terms are presented. “Privacy” is discussed in the next section, i.e. section 2.3.

CONFIDENTIALITY: The International Organization for Standardization (ISO) has defined “confidentiality” as “ensuring that information is accessible only to those authorized to have access” [ISO05]. This means that information should only be disclosed to “the right people” and to nobody else. Confidentiality can be ensured by using proper access control and by using cryptosystems for securing data transmissions. Confidentiality is related to data privacy, i.e. limiting access to individuals' personal information.

INTEGRITY: The International Organization for Standardization (ISO) has defined “integrity” as “safeguarding the accuracy and completeness of information and processing methods” [ISO05]. Regarding information systems, this means that systems cannot be tampered with to provoke improper operation. Attackers or coincidence should not be able to provide false data or alter valid data unnoticed or bring a system in an unwanted state by any means. Within communications, integrity therefore includes that the origin of data is proved and that the data has not been changed unnoticed upon transmission – whether by accident or by hostile activity.

AVAILABILITY: The International Organization for Standardization (ISO) has defined “availability” as “ensuring that authorized users have access to information and associated assets when required” [ISO05]. This means that systems and infrastructure need to be available to provide service whenever a legitimate user requests a service. Data needs to be readily available upon request of legitimate users. Like integrity, availability can be affected by accident (e.g. malfunctioning devices or transmission errors) or malign influence.

Besides the stated categories, there are some other properties for secure systems that are of relevance in practice. The *complexity* of a system should be as low as possible. This makes a system easier to create and to maintain. If the operational sequences within a system are easy to understand, the operation of a system is more transparent to a user. Such a better *transparency* gives the user a feeling of control over the systems which increases the user's trust into the system's operation. If a system interfaces with users directly, also the *usability* of a system is important. Technically, usability is of no relevance, but without good usability users try to avoid security features due to their complexity or do not make use of a system at all (see e.g. [WT99]).

2.2.2 Safeguards

The term “risk” expresses the likelihood of harm arising from threats [Cla01]. The risk associated with the operation of a system should obviously be as low as possible. To get a quantitative means to evaluate risks, one can use the widely known risk equation to get a monetary risk value: $risk = threat \cdot vulnerability \cdot cost$ (see e.g. [FEMA03]). Threats are circumstances "that could result in harm to the entity" [Cla01] and "threat" in the equation is the frequency in which such adverse circumstances occur. "vulnerability" is "the susceptibility of an entity to a threat, in the form of a weakness that may permit a threatening event to give rise to harm" [Cla01] and "cost" is the monetary loss or damage that would result if the security of a system failed due to a threatening event.

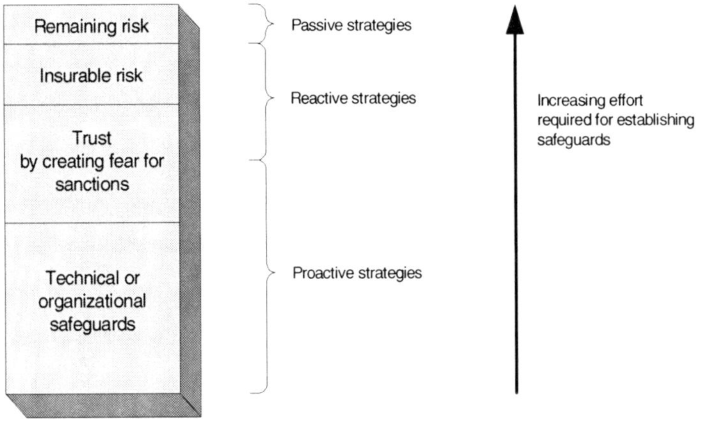

Fig. 2.5. Methods for coping with risk

There are different methods for coping with risk that is introduced by insecure systems. These are shown in figure 2.5. At first, risk can be avoided or lowered. Threats can be eliminated, or at least the vulnerability can be significantly decreased. Methods can be the implementation of technical safeguards (hardware, software, and networks), e.g. applying protection by cryptographic means, or organizational safeguards that inhibit threats to occur by using the system only within proper structures and processes. This is a proactive risk management strategy. Economics tells us that the effort required for implementing such safeguards needs to be lower than the monetary risk imposed by a system without that safeguards. From the view of an attacker, it makes sense to perform an attack if there is an incentive, i.e. a high enough profit to take. There is an incentive if the expected profit from a successful attack is higher than the efforts required bypassing the safeguards.

A second measure that can be proactive but also reactive is not to install safeguards that eliminate threats or decrease vulnerability. Harm is prevented in such a way that possibilities are created to detect malicious actions and to identify the attacker. Therewith hostile actions can be effectively sanctioned. Thus, an attacker is able to do harm but will not do so because of the impending sanctions as long as the incentives for the attack do not outweigh the expected sanctions. In the case that a threatening event occurs anyhow, one is able to detect the malicious actions and can penalize the identified attacker. For this case, there should exist an appropriate way to recover from the occurred threatening event.

A third measure for risk management that is also reactive is insurances. Here the risk is transferred to another entity. If there are no or insufficient safeguards and harm is done, the insurance will answer for the occurring damage. Insurances can be attractive if establishing additional safeguards is too expensive and if a threatening event is of low probability but of high monetary impact. Insurability of risks regarding current technologies and developments like e-commerce is under research [Grz02].

The remaining risk that is not managed by any other means needs to be tolerated. Accepting a risk is an option if no additional safeguards can be implemented with reasonable cost and there is no insurance available for the threatening events that could occur.

Within this book, technical or organizational safeguards shall be established to cope with the risk that is associated to the threats that RFID systems are exposed to. If such safeguards are not feasible for any reason, at least the possibility to detect malicious actions and to identify attackers shall be set up to enable sanctions against such actions. Note that imposing sanctions alone without a possibility to identify attackers is not sufficient. For instance, sending unsolicited email ("SPAM") is prohibited by law, but it is nevertheless seldom possible to take court action or at least to stop the spammers.

2.2.3 Security by Design

Systems should always be designed with security in mind, i.e. security should not be an afterthought but an important design goal [HM05]. With a transparent, easily maintainable system architecture that makes coping with complexity well possible, the probability of vulnerabilities can be kept low. There are additional design principles like e.g. building "multiple barriers" that help with designing secure systems.

The RFID systems and communication protocols that are proposed within this book are designed with security as a design goal. The resource scarce environment of RFID tags does not make is possible to add safeguards afterwards (like for instance adding SSL/TLS [DA99] to an otherwise insecure service) but requires security to be an integral part of the system design.

2.2.4 Security Summary

In this section the term "security" has been discussed and some guidelines for building secure systems have been gathered. Making RFID systems and communication protocols secure is a goal within this book. To achieve it, appropriate safeguards or at least suitable detection mechanisms will be created. It is beneficial to make security part of each element in the system architecture. Within RFID systems, not only the technical properties are relevant and thus considered but also secondary properties like usability and transparency that are important for the users' perception of the systems. These secondary properties become even more important with regard to privacy, which is considered in the next section.

2.3 Privacy

Privacy is a central element within this book and will thus be analyzed in greater detail in the following. The ubiquitous computing systems that are emerging today are intended to be context sensing so that they can derive the user's intention to better support him. Gathering the user's context requires vast amounts of data representing the user's actions and behavior. This can obviously lead to privacy implications. "In fact, plenty of today's pervasive applications could present potential threats to privacy and liberty" [Sto03]. This is a huge problem since privacy is regarded as a prerequisite for a "humane and livable information society" [Roß05].

When people claim that privacy should be protected, it is usually unclear what they mean precisely. Articulating the possible harm by missing privacy is difficult, and often it is even unclear what privacy is. This is well expressed in newspaper reports about privacy violations, but such reports are often not objective. "The typical privacy article rests its case on an appeal to its reader's intuitions and anxieties about the evils of privacy violations" [Whi04]. "Commentators often fail to translate our instincts into a reasoned articulable account of why such a privacy problem is harmful" [Sol06]. When even the reporting about privacy appears to be difficult, modeling architectures and system designs for technical solutions becomes more than sophisticated.

The problem that a good understanding of privacy for implementing technical solutions is required is realized today and has found its way into many publications. For instance, Langheinrich wrote much about that topic in several papers and also in his dissertation [Lan05], for instance, that it is "crucial to understand when it is exactly when people feel that their privacy has been invaded" [Lan02].

The following disquisition on privacy aims to summarize the interesting topic "privacy" with the intention of deriving guidelines for modeling technical systems supporting the actual needs of users regarding privacy.

2.3.1 Historical Overview

Privacy has deep roots in history. Mostly in the form of a right to solitude, it can be traced back into early Hebrew culture, Classical Greece, and ancient China. Furthermore, references related to privacy are already found in the Bible [GILC98]. But until its codification in modern law centuries passed.

In 1361, the Justices of the Peace Act in England provided for the arrest of peeping toms and eavesdroppers. This was the first known trace of legislation regarding privacy.

Another aspect of privacy was covered by Sir William Pitt, Earl of Chatham, in 1763: "The poorest man may in his cottage bid defiance to all the force of the Crown. It may be frail; its roof may shake; the wind may blow through it; the storms may enter; the rain may enter – but the King of England may not enter; all his forces dare not cross the threshold of the ruined tenement" [Uro03]. This expressed the right of an Englishman to be secure in his home and introduced the understanding of one's home as a protected place: "My home is my castle."

With the rise of daily newspapers another threat to privacy emerged. With them, life of people could be made public and their repudiation be affected negatively. This concern found expression in the law of France as "Loi Relative à la Presse" in 1868. This illustrates the changing notion of privacy due to emergence of new technology like newspapers or the telephone.

In 1890 a law review article titled "The Right to Privacy" written by Warren and Brandeis was published [WB90]. This article is seminal because it introduced a definition of privacy that even nowadays is in use: Privacy as the right to be let alone.

In "The Right to Privacy", the idea of one's "full protection in person and in property" that already became manifest in the right to life, liberty, and the right to property was broadened in scope. It is argued that due to political, social, and economic changes new demands of society arose and that consequently those legal rights were no longer only recognized as remedy for physical interference but also had to account for the "spiritual nature", i.e. one's feelings and intellect. Because of that, the right to life was given the wider meaning of a right to enjoy life which included the right to be let alone.

In 1948, the United Nations proclaimed in its "Universal Declaration of Human Rights" [UN48] that "no one shall be subjected to arbitrary interference with his privacy,... Everyone has the right to the protection of law against such interference or attacks." as a basic human right.

This was an important step, and with it, the concept of privacy found its way into the legislation of many countries. A comprehensive overview of the development of legislation regarding privacy laws in different countries, the topic filling a whole book, can be found in [EPIC04].

2.3.2 Defining Privacy

As it is represented in the historical development, the perception of what privacy actually is changed along with technological developments and ongoing changes in society. Furthermore, there is a personal perception of privacy. The actual meaning of privacy as well as its value is subjective and differs between different people. Because of that, an accurate general definition does not exist and one must stick to circumscriptions that are subject to different interpretations.

There is neither a commonly agreed-upon definition of privacy nor anything that comes near to that: "Privacy is a fundamental human right. It underpins human dignity and other values such as freedom of association and freedom of speech. It has become one of the most important human rights of modern age.", "Of all the human rights ... privacy is perhaps the most difficult to define" [EPIC04]. There is a different understanding of privacy over time, within different cultures, and even between single individuals. This does not only lead to problems in the system design of "privacy respecting" systems but also in law: " 'privacy' means so many things to so many different people that it has lost any precise legal connotation that it might once have had" (reflected in [Sol06]).

Brandeis' already cited "The right to be let alone" is a definition that is often heard of but that is too abstract for deriving system properties. It already shows that privacy is related to individuality which is a prerequisite for the need for privacy: "without a sense of individuality, there can be no perception of a need for privacy" [Uro03]. This means that the need for privacy comes from the need for personal autonomy. This autonomy has to be given from society: "The need for privacy is a socially created need. Without society there would be no need for privacy" [BM87].

A longer, more concrete definition is the following: Privacy is "the desire by each of us for physical space where we can be free of interruption, intrusion, embarrassment, or accountability and the attempt to control the time and manner of disclosures of personal information about ourselves" [Smi00]. This definition shows that privacy has different aspects like seclusion, i.e. the desire to be left alone, and autonomy, i.e. the ability to act freely. The latter manifests in particular in the ability to control the dissemination and use of one's personal information, i.e. the control which information should be communicated to and used by others.

A categorization of aspects of privacy is presented in [EPIC04] and sometimes called "facets of privacy": *Territorial Privacy* (concerns the entry to personal spaces), *Bodily Privacy* (concerns the protection of people's body, e.g. against invasive procedures like cavity searches), *Communication Privacy* (covers the privacy of communications like telephone or email), and *Information Privacy* (controlling the collection and handling of personal data). These four categories highlight the diversity of people's privacy interests.

An often cited definition that can for instance be found in the Internet draft "draft-haddad-alien-privacy-terminology-01" and also in papers by privacy researchers like Langheinrich comes from Ryan: "Privacy is the claim of individuals, groups and institutions to determine for themselves, when, how, and to what extent information about them is communicated to others" [Rya67]. This definition focuses solely on personal data and does not consider the other aspects. Such personal data can be of various forms like one's identity or identifying data (id numbers, genetics, biometrics), data linked directly to an individual (like address, date of birth etc.), and context information (location and activity).

The variety of definitions shows that privacy is a complex topic. The definitions express what aspects are important to people. But they do not show *why* there is a desire to privacy and do not give a rating which aspects are essential and which are of minor importance. Therefore, the next subsection aims at getting a better understanding by discussing the importance of privacy.

2.3.3 The Importance of Privacy

The importance of privacy becomes clear when looking at the extremes: These are *no privacy* on the hand and *total privacy* on the other.

No privacy

Imagine a world without privacy. Everybody could watch your personal spaces like your bathroom or bedroom. Everybody could analyze your genes. Everybody could read your letters and your email and could listen to your telephone calls. Everybody could access any data that is stored about you which would be a huge amount: About your relatives and friends, your health status, your bank account and transactions, your insurances, and anything else – each and every interaction of you with the outside world. In the near future, even more data than today will be collected and stored: Thus, everybody would be able to retrieve your behavior and actions from childhood until your current age with all the conversations you had, perhaps even with video.

As confidentiality is also an aspect of privacy, there would be no confidentiality possible any more: Everybody could get any information he wants. There could be no secrets any more. Only your thoughts would be private, because there would be no practical and powerful means to gather more than activity information from the brain – but perhaps there will come a time when even your thoughts would no longer be private. Passwords and PINs would no longer work because they could not be held private any more. Even ordinary key locks will no longer give security since everybody could access the information required to forge them. Even your personal diary would no longer be private.

This scenario is much more than a Big-Brother show. There would be no spaces for retreats any more. Nobody would be able to keep any secrets, neither individuals nor companies. Another problem of insufficient privacy is that people are under pressure to act in a "normal" way. But the dignity of man demands that he can decide and act freely and self-determined (see [Mat04]).

The consequence is that privacy needs to be regarded as basically important so that it does not make sense to abolish privacy. "A free and democratic society requires respect for the autonomy of individuals, and limits on the power of both state and private organizations to intrude on that autonomy. Privacy is a value which underpins human dignity and other key values such as freedom of association and freedom of speech" [APCC94].

Total privacy

Total privacy would mean total freedom to decide about ones personal space, affairs, and information. This extreme also does not work because it puts the interest of a single individual over legitimate interests of the society. The freedom of somebody ends where the freedom of somebody else begins: "No one ought to harm another in his Life, Health, Liberty, or Possessions." (John Locke, English philosopher, 1632-1704).

The reason for this is that "there are certainly times when people should be held accountable for their activities in private" [AC03]. A good example why society has an interest to sanction activities performed in private is obviously criminal activities.

Besides this argument, another problem would be the implementation of total privacy: There could be no public interaction between people. The problem increases with the use of technical equipment: For instance, if someone wants to take a photo, this could harm the rights of others. One would need to ask for permission. Obviously this is not feasible, if there are many people. But it is also impossible if there are no people on the picture: The picture could show personal territory of somebody. A reproduction of this personal territory would harm privacy. But the photographer would have no possibility to ask for permission because there is perhaps no way to identify the person to which the territory belongs. There are also many other examples that show that the use of technical equipment and even normal social interactions would be rendered impossible (How to avoid an accidental meeting of persons if there is no way to ensure that it does not happen due to privacy reasons?).

The conclusion is that total privacy is a state that is neither wanted nor possible. It would trim the rights of society, cause implementation problems, and even render social interaction impossible without limiting other fundamental rights.

2.3.4 Privacy Today

Today, privacy is a fundamental but not an absolute human right that is codified in the constitutions of many countries and in many international treaties. Examples are the already cited "Universal Declaration of Human Rights" of the United Nations [UN48] or the International Covenant on Civil and Political Rights [HRC66].

In the United States, privacy is only implicitly recognized in the constitution: The Fourth Amendment says: "The right of the people to be secure in their persons, houses, papers, and effects, against unreasonable searches and seizures, shall not be violated...".

The magnitude of laws concerning the protection of privacy expresses what a worthwhile and highly valued right privacy actually is. The already discussed description "the right to be let alone" introduced by Supreme Court Justice Louis Brandeis in 1890 [WB90] is until today the most widespread circumscription of the concept of privacy, but the implementation in law differs among countries. A summary giving a good overview can be found in [EPIC04].

The central question in law regarding privacy is the matter how far society may intrude into a person's life or affairs and what the rights of an individual for privacy protection are. Here, the rights of the individual have to be brought into a proper balance with the rights of the community.

2.3.5 Current Development

Technological developments produce new threats to privacy. Today's information technology has a vast capacity to collect and analyze information on individuals. Thus, the potential for data abuse and privacy invasions is increasing accordingly. Relevant current technology trends are miniaturization, embedding, networking, ubiquity, and context awareness [BSI06]. There is a variety of small, cheap, and numerous devices that increase the amount of information created by each individual. Large storage space and high processing power are enablers for large databases that are able to store and analyze this data. The Internet and mobility technology link the devices and databases and remove geographical limitations to the flow of data. Service-oriented architecture removes the technological barriers between individual systems.

Erosion of privacy

The stated technological development leads to vast amounts of data that threatens privacy of the individuals because companies as well as governments find an interest in using that data.

Network addresses of personal devices like notebooks or mobile phones can be seen as pseudonyms for the device user. Examples are MAC addresses, IP addresses, Ethernet MAC addresses of wireless devices, Bluetooth hardware addresses, and the International Mobile Equipment Identifiers (IMEIs) of mobile phones. For the possible abuses, sometimes even new words have been created, e.g. "bluetracking" [McF05]. Mobile phone providers can position their users using the address (IMEI) of the mobile devices. This feature is used technically to enable roaming as well as for the provision of location based services but can even be used for tracking users.

But device users are not only logged and tracked in computer networks but also in other infrastructures: Number plates or devices for toll collection or special GPS devices can be and are used to track vehicles. Today, the police checks number plates for stolen cars and invalid insurance, data is collected for traffic billing purposes, and truckage companies as well as rental car companies use GPS devices for tracking their vehicles. Enhanced versions of such tracking devices are currently introduced by some insurers in Great Britain to enable a "pay as you drive" insurance tariff.

Companies have a huge interest in collecting data about their customers, e.g. for marketing purposes. Thus, they collect a vast amount of data. An example is the data generated by using credit, debit, and store cards. In a German newspaper article ("Die Zeit: Wir werden täglich ausgespäht" [The Times: We are spied out each day]), Martin Franssen, an ex corporate consultant of American Express, is cited about the extensive data shadows the users leave behind: "The large providers are able to conduct attitude surveys. They can trace back for ten years where the user using his card regularly lingers, where he sleeps, whether he travels often, whether he drinks much, or whether he makes women too expensive gifts. In the bedroom, the Germans are afraid of Peeping Toms; at the sales counter they become exhibitionists." The data collected is used to prevent fraud and sometimes for other purposes like marketing, too. Sales companies also perform data mining to be able to perform personalized marketing. Publicly available data is collected and afterwards sold by professional address and data brokers, too. There are also central databases that are maintained for checking the creditworthiness of people and companies and for giving references about past problems like belated paying. In Germany, the "Schufa", "Creditreform", and insurance companies operate such databases. These databases are a "pillary" for non-conforming behavior, and it is often difficult for an individual to get to know about wrong entries and to get them changed or deleted afterwards.

But not only companies collect vast amounts of data. Governments increasingly try to get access to data and also create databases and other installations on their own, too. Examples are email and Internet surveillance, DNA analysis and gene databases, biometrical data, RFIDs e.g. in passports, health insurance cards, video surveillance on public places and public transport. In Germany, banking confidentiality has been considerably lowered recently giving many government agencies access to data about bank accounts. The police get publicity for using malware for the surveillance of suspects. The United States installed the no-flight database (and its successor) and dictated other countries to use passports with biometrical data to allow entry without visa.

Besides the stated examples, there are many others where technology threatens privacy. For instance, the enforcement of registration of software and regularly checks for genuineness, digital rights management (DRM) systems (see [Sur06] for privacy implications), the copy counterfeit identification system in printers that even earned a Big-Brother award, or the planned enciphering of TV channels with the goal of addressing individual viewers in the future.

Basic assessment

Summed up, very much data are collected: practically everywhere and always, often unobtrusively, sometimes even invisibly, detailed and individually. It is done "in the name of law enforcement, security, cost-saving, and convenience" [KC04]. Rivest, one of the inventors of the cryptosystem RSA, calls the current development "reversal of defaults". He says that "what was once private is now public; what was once hard to copy is now trivial to duplicate; what was once easily forgotten is now stored forever" [Riv01].

Much of the collected data is used for reasonable purposes, for instance increasing comfort of people, but it could also be used against the interests of people. "You can make a chunk of wood into a violin or a very effective club" [Sto03]. Schneier speaks about "dual use technologies" in this context (see [Sch03]): Almost all technologies can be used for good and evil. There is a trade-off: If the good uses outweigh the evil uses, the technology should be used and one needs to find a means to deal with the evil uses in some way.

If one assesses the current development from a critical perspective, privacy protection in most countries seems to have failed. Government agencies and companies have gained much more power compared to a decade ago. Data protection commissaries that are present in many countries and are also mandatory in Germany for companies exceeding a certain size do not have a high impact any more. Compared to the loss of privacy on a big scale, their influence gets low and their work shifts into less important areas which makes them appear pettifogging and makes data security appear to be an obstacle for useful applications [Mer06].

Many people are afraid of the increased power of government agencies and their intrusion into affairs that have been private before. Surveillance and data collection regarding single suspects makes way for surveillance of everybody and afterwards deleting the data that is not required. In "Your vanishing privacy", a Star Tribune article, Schneier speaks of "wholesale surveillance". There is no longer a "follow that car, watch that person, listen on this phone conversation" but an "eavesdrop on every phone call".

Much of the new surveillance installations are justified by the danger of terrorism. But the effect is questionable: If the surveillance increases, there is relocation to spaces that are not under surveillance. The effect is often not increased safety but only decreased freedom (see [KC04] and "Why Data Mining Won't Stop Terror", http://www.wired.com/news/columns/1,70357-0.html). Benjamin Franklin who helped draft the Declaration of Independence said "Those who would give up essential Liberty, to purchase a little temporary Safety, deserve neither Liberty nor Safety."

The data collection of companies is also sometimes questionable: The "pay as you drive" tariff model proposed by some car insurance companies makes the amount invoiced dependent on parking sites, overtaking, speed, street conditions, daytime etc. (see e.g. [Mat04]). The privacy implications are enormous, but on the other hand, the costs better correspond to the actual risks. The tariffs are thus fair for the customers but much more complex, and the average insurance fee will stay the same as long as the general risk is not lowered by intimidation of the drivers. Thus, customers should always ask whether they need a development or they should better refuse it.

Development characteristics

In the following, some characteristics of privacy and the associated problems caused by the recent and coming development are explained.

DYNAMIC CHARACTER: In the subsection about the historical development and in the ones about the privacy definition and the state and development of privacy today, it became obvious that privacy is not something static: "The public and the private involve multiple meanings over time and across cultures, contexts, kinds of persons and social categories" [Mar01]. The greatest changes appeared over time: "Various historical changes have brought about a change in perspective of our privacy needs" [Lan01]. Most changes are caused or at least initiated by the already explained technological changes like in microelectronics. The further technological progress will probably intensify the challenges regarding privacy protection.

LINKING OF DATA: Single data collections and single databases are often not a threat to privacy. But the combination of such data could become large a threat (see e.g. [Mat04]). Single aspects about an individual can be put together to comprehensive information about that individual. Much data emerges by the use of devices and by performing certain activities. In sum it provides detailed information about a person's behavior and activities.

NO OPT-OUT: The problem of linked data will become more and more evident if the number of devices in our environment increases. Ubiquitous computing will bring small and unobtrusive devices, which can produce data shadows of people, into our daily life. There is a major difference to former technology: In the disability to opt-out. For example, if somebody dislikes mobile phones he can simply decide not to use such devices. But when the devices become ubiquitous around us there is no longer an option for choice. For instance, there will be no way to decide not to use RFID technology for the individual when such devices get affixed to bank notes, id cards, and every item that one can buy in a store.

OPT-OUT DIFFICULTIES: Even if an opt-out is theoretically possible, there can be problems for the individual. One problem is that the decision to opt-out could cause that much inconvenience and additional work (e.g. postal mail of a check instead of a credit card payment) that there is no real option. Another problem is that an opt-out could look suspicious. Often, a denial to give away data leads to suspicion. For instance, if a person does not want to take part in a voluntary DNA analysis after a crime has been committed, this could be interpreted such that the person hides something.

LACK OF EVIDENCE BY OPT-OUT: The previously stated scenario can even be drawn further: Perhaps the presumption of innocence will not hold any more, and not guilt needs to be proven anymore but innocence. Mattern gave an example in [Mat04]: "If the accused had not wanted to conceal anything, he would not have switched off his remotely localizable identification device in the critical moment" (translated from German).

LOSS OF EPHEMERAL COMMUNICATION: In [Sch06a] and [Sch06b], Schneier warned for another effect of the current development: "We are rapidly turning into a society where our intimate conversations can be saved and made public later. This represents an enormous loss of freedom and liberty.", "The moral is clear: If you type it and send it, prepare to explain it in public later." Here, today's ability to store large amounts of data creates a technical possibility that produces privacy implications.

This shows that there are some characteristics in the current development that should make people think about their attitude towards privacy. However, the perception of people regarding privacy threats often does not conform to the actual ones which will be shown in the next subsection.

2.3.6 Perception of Privacy

Technical safeguards for protecting privacy need to follow the actual requirements of the users. In the previous subsections, the nature of privacy has been explored and the current development has been shown and reasoned. These considerations did not include the perceptions of people. Thus, this subsection discusses people's perception of privacy.

Today, data is stored in that many databases by that many companies and institutions that nobody knows what data is stored about him and who owns that data for what purpose. People are giving away more information than they probably notice [KC04]. The reason is that everybody leaves data shadows, for instance when telephoning, when surfing the Internet, or by video surveillance in banks, other buildings or on public places. The resolution of the information about an individual that can be aggregated using the collected data altogether is already very high.

There is often a huge difference between perceived and actual risk: "we over-react to intentional actions and under-react to accidents", "we over-react to things that offend our morals", "we over-react to immediate threats and under-react to long-term threats", "we under-react to changes that occur slowly and over time" ([Sch03], pp. 26-27, and focused in [Sch06d] as well as in [Gil06]). Thus, people's privacy perception is not objective so that the perception of the threats for their privacy resulting from the current development is also not objective.

As mentioned, the development is often not perceived by people clearly. The result is that only few resist the development and try to opt-out. Many people do not know how technical devices work and what the risks to privacy are. After explaining in detail what might happen, people get much more sensitive to the threats. But a problem is that warnings about threats often do not have a positive result in the long-term: Warnings about threats that do not lead to harm within reasonable time are perceived more and more as unsubstantiated. The harm also needs to be recognized e.g. by exposure in media.

Another reason why the development is not perceived by many people is that it occurs step-by-step. People seem to get used to giving away their data, and usually only the advantages for doing so are communicated to them. The development is not realized because many people do neither have the time nor the notion to get informed and thus orientate on the behavior of the masses. Nevertheless, people's concern over privacy violations is steadily increasing and "now greater than at any time in recent history" [GILC98].

Individual privacy expectations

Each individual has a certain expectation regarding privacy. This expectation is shaped by societal norms that are common in the individual's environment (like era and culture) and values of the individual. This is depicted in figure 2.6 on the left side.

A person's individual expectation is the measurement scale for privacy violations: Under all the things that could happen ("possible occurrences" in figure 2.6), the actual event is rated using the individual expectation. If occurrence and expectation do not match, the event is perceived as being accompanied by a privacy violation.

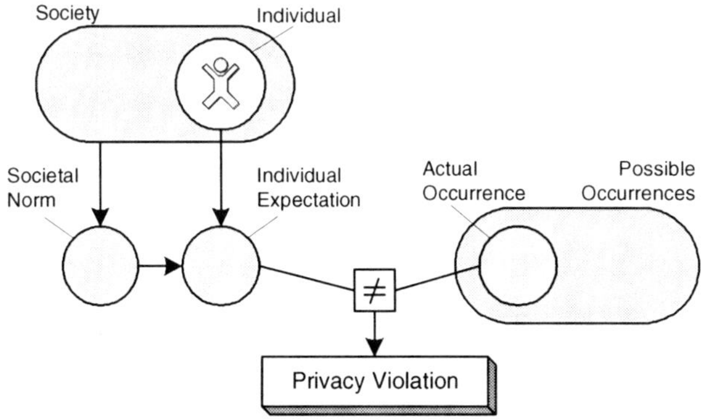

Fig. 2.6. Privacy violations depend on individual expectations

Several researchers tried to model the relevant aspects for considering an event privacy violating or not. Jacobs found that the privacy perception regarding sensor devices depends on input stimulus, location of input stimulus, location of sensing, and granularity [JA03].

Marx introduced the model of "border crossings" [Mar01]: "Central to our acceptance or sense of outrage with respect to surveillance ... are the implications for crossing personal borders". He distinguished different kinds of such personal borders: "Natural" borders are restrictions to senses like e.g. walls or letter envelopes. Tools that extend the senses can lead to a border crossing so that they require notice or special permission, i.e. consent. "Social" borders are expectations of social roles and behavior. For instance, there are certain expectations regarding doctors or friends. "Spatial" or "temporal" borders separate different periods or aspects of life. Different aspects of one's personal biography, especially concerning different locations, are expected to be isolated from each other. There is also an assumption that social interaction like communication is only transitory and thus ephemeral. Here, a border crossing would be performed by a hidden audio or video recording.

Models like the one based on "border crossings" help to understand when privacy violations occur. Based on this understanding, one can try to find an answer to the question *why* privacy is important to people, i.e. what the source of privacy expectations is. In the following, some important reasons for our privacy needs are presented.

Personal freedom and autonomy

Humans do not want to be under surveillance and also do not want to have a feeling of surveillance. They want to have the freedom to decide themselves on themselves how to act and live.

This is supported by the following citation: "If one has an enduring feeling of being under surveillance, one feels stressed and thus is constricted in one's mental integrity, freedom of action, and freedom of reasoned assessment." (translated from

[Sur05]). Dueck criticizes that people are watched perpetually, too: "The computer guards us" (translated from [Due06b]).

Another statement is the following: "The classic example is law enforcement officials having too much power, which can alter the way people engage in their activities. People might be chilled in their behavior, making them less likely to attend political rallies or criticize popular views" ([Sol06], page 7). This is what is also known from TV shows like "Big Brother". Personal behavior becomes watched and possibly sanctioned by others. This leads to a change in behavior. This problem is also stated by Mattern in the context of ubiquitous computing: "The technical possibilities of ubiquitous and pervasive computing could lead in particular cases to a constriction of the freedom of opinion, making self-determined actions more difficult, and to loss of control" [Mat04]. The next paragraph covers why people want to maintain control.

Control and non-abuse

"People are willing to share all sorts of information, as long as they are in control" [Sch06c]. The reason is that the information can be abused. This means that it is alright to share information and to relinquish privacy as long as one has enough trust that the information is not used against one's interests.

If no information is shared, there is no need to ensure that stored information is correct. Information that is stored about people is given more and more evidence. Erroneous data can lead to severe problems for people. "What happens if data in an IT system can back a fact with numbers and other records and on the other side there is 'just' the statement of the one concerned?" (translated from [Sur05]).

There is also no need to explain oneself about things nobody knows if no information is shared. This is important for our daily lives: There are many laws and rules that have been set-up with a certain intention. The rules leave room for interpretation. Moreover, rules are quite often not suited for a current problem, i.e. the intentions are not met. Then it makes sense to change the rules or to disobey them. As a change is often too complicated or would make the body of rules too complex, a reasoned disobeying of rules is the only appropriate solution in practice. Neglecting rules sounds bad but is often helpful and positive. It enables people to be merciful and enables an appropriate behavior if there are occurrences nobody could image when the rules were set-up. Dueck deals in his column in the German "Informatik Spektrum" magazine with people within the "forces of the system" (i.e. an organization in his context). He argues that it is important to have people's intelligence within the system and that it is not possible to automate everything with a set of rules [Due06a]. But "creative freedom" is only possible with ease as long as there is privacy to allow it without the need to explain oneself.

Surveillance society

People fear that a surveillance society could be established that reduces or even extinguishes personal freedom and the control over their data. The technical infrastructure we have today could be abused by a totalitarian government. Only laws are in the way of abusing the systems. Legislation has changed "post September 11th" in many countries giving government agencies more power.

Sometimes it is argued that privacy makes it more difficult to catch criminals. That is true but leaves some other important points behind. Lowering the protection targets does not only hit criminals but also freedom and dignity of all honest people. For catching some more criminals than before, all the citizens are under surveillance today. Even more, a lowering of privacy protection could help criminals: "As more information about us is collected, correlated, and sold, it becomes easier for criminals to get their hands on the data they need to commit fraud" [Sch05]. Technology can prove "fatal to anyone 'of interest' to a regime" [GILC98].

" 'We are building the infrastructure for totalitarian control,' says Deborah Johnson. A professor of applied ethics at the University of Virginia,... 'Right now, people are not afraid of it because it is not being built by the government. It's being built by the market and by commercial interests, but once it is all set up in place, it will only take a slight shift in political ideology for it to be used in other ways" ' [Sto03]. The problem is also seen by the German federal commissary for data protection: "It is legitimate to use the technological developments,... At the same time, technical control systems and surveillance infrastructures are upgraded,... whose lawful and data protection aware operation cannot be controlled ultimately any more". He sees a "threat for informational self-determination", "which is at first not recognized by the concerned people and within the societal discourse" (translated from [BfD05]).

Balance of power

People feel concerned or embarrassed if others know more about them than vice versa. Such imbalances in knowledge lead to imbalances in power that make people feel uncomfortable or even scared.

An example of such an imbalance is when one person in a room is naked and all other persons in the room are dressed. The state of being naked is (for most people) not a problem as long as the others are naked, too, e.g. in a sauna. But if there is an asymmetry, it becomes a problem: The naked person amongst the others appears to be special and draws attention. The person might even feel to be committed.

Similar imbalances can appear with data. An extreme case is when people are able to watch others but these watchers are not watched themselves. Then the watchers are a kind of peeping toms that act in the background without being noticed or harmed for their behavior. "Imbalances in power can also be risk-enhancing, in that they increase the risk of abuses of power" ([Sol06], page 7).

Because of the problems that are caused by imbalances of knowledge and power, there are people like Brin and Goldstein that ask for a "transparent society" (see [Bri99] and [Gol04]). The train of thoughts is that the deployment of technology and infrastructures that can be used as a means for surveillance, e.g. by a totalitarian government, cannot be stopped. As a consequence, one cannot stop the watchers but can make the watchers being watched, too. This means that an information balance should be created by making information available to anybody and not only to the government. Anybody should be able to get information about you, and in turn, you should get to know that and be able to get corresponding information about the other person.

Langheinrich also explains the idea: "If everybody has access to the same information, it ceases to be a weapon in the hands of the well-informed. Only when the watchers are being watched, all information they hold about me is equally worth the information I hold about them. Eventually, new forms of social interaction will evolve that are built upon these symmetrical information assets" [Lan01]. The idea of a transparent society requires some changes in our pattern of thinking but is very interesting. Unfortunately, the approach also has some problems that need to be solved but are not discussed exhaustively in the stated publications: What happens to prominent people or people that are special in another way? Can they live unattended any more? Or will information about them become uninteresting because everybody has stories? What about stalkers? How shall the notification about access to one's data be implemented? Such questions need to be answered. However, a detailed discussion is beyond the scope of this overview. But the general idea is important within the remainder of this book.

2.3.7 Regulation Approaches

There are different approaches for implementing privacy protection schemes: Self-regulation, legal regulation, and technical regulation. These will be presented in the following.

SELF-REGULATION: The idea is that companies protect privacy voluntarily without being required to do so by legislation and albeit having the technical infrastructure for collecting and using much more data. There are two kinds of incentives for a company to restrict themselves: *social norms* and the *market*.

Self-regulation approaches often appear to be no more than image campaigns. Thus, self-regulation proved to be disappointing in the past. The incentive for privacy protection is often too low for companies so that problems regarding adequacy and enforcement occur: The privacy protection is weak, and there is a lack of enforcement (see [EPIC04], page 4).

LEGAL REGULATION: Here, privacy protection is demanded by *law*. The technical infrastructures might allow collecting and using data, but it may not be done due to legislative restrictions.

The experiences with legal regulation alone are not encouraging: Mere presence of a law does not provide adequate protection. To detect illegal actions and sanction these is often not possible, i.e. it is difficult to enforce the laws.

A good example of the lack of law enforcement is the SPAM problem: It is prohibited to send unsolicited email but it is not possible to enforce such laws. Similar problems are illegal music downloads: Albeit the sharing of copyrighted material is prohibited, it is very difficult to prevent this. Langheinrich hits the point: "It is important to remember that laws can only work together with the social and technical reality, not against them" [Lan01].

TECHNICAL REGULATION: The third regulation approach is the technical regulation. This means that there are technical safeguards that ensure privacy protection within the *architecture* of systems, i.e. "privacy by design" [Lan01]. This kind of privacy protection is a strong one, but often it needs to be applied in the design phase of systems and cannot be provided as an add-on [HM05]. Technical regulation is more than just a more secure processing of data: "It is important to distinguish between genuine privacy enhancing technologies and data security technologies that seek to render processing safe but not to reduce the disclosure and processing of identifiable data" [Bur97].

Regulation in practice

In practice, only a combination of technical and legislative regulation can provide a high privacy protection level. This thesis is also supported by Rossnagel, e.g. in [Roß01]. There need to be technical safeguards [Roß05] so that abusing data is not easily possible any more. Additionally, there need to be laws that demand the implementation of such technical safeguards, that sanction bypassing effective technical safeguards, and that cover areas in which technical safeguards are not appropriately implementable. Regulation by the market also works in special cases; but regulation by social norms works well only between individuals and not between individuals and companies due to the laws of the market.

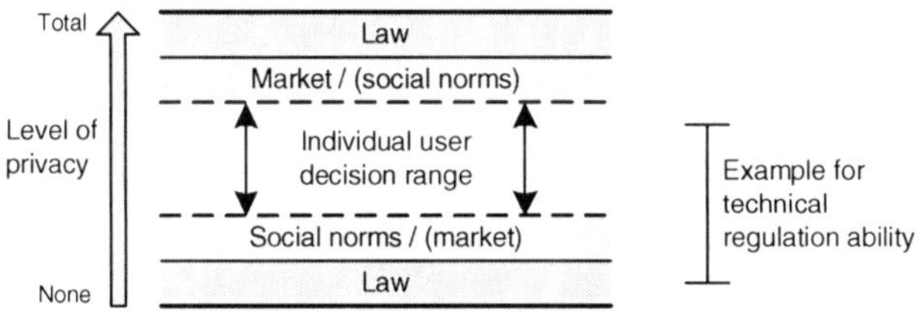

Fig. 2.7. Regulation approaches and privacy

In figure 2.7 an example of the cooperation of different regulation approaches is depicted. Law gives boundaries for the minimum and maximum privacy level between which the user can choose: Law requires some privacy protection but also limits privacy. Additional boundaries are introduced by social norms and the market. Ideally, laws and the requirements resulting by current social norms should match, but there might be a difference in practice. Within the given constraints, the user should be able to choose the level of privacy he wishes.

The architecture of technical systems gives hard limits for the possible privacy level. An example is drawn into figure 2.7. There, the technical system does not fulfill the privacy requirements, because the user might want to choose a privacy level that the system is not able to offer. This is especially bad if the user cannot opt-out from using the technical system. The example also shows that the technical systems might offer possibilities that are not wanted due to legal, social or market constraints. If a user chooses to use a technical system in such a way, there should be at least a means to detect such abuse so that one becomes able to sanction the misbehavior.

In practice, implementing technical safeguards is not an easy task because the goals are often not defined clearly. As already discussed previously, the perception of privacy depends on many aspects, e.g. the individual perception. A balance between different privacy demands and the effort that is required for technical safeguards needs to be found. Usually, there are conflicting interests between different parties (see [Roß05]): Citizens/consumers have other interests than companies or governments so that an appropriate balance needs to be found – ideally before system design. Interesting questions are, for example, which "backdoors" shall be implemented for detecting abuse and monitoring the systems and which data shall be given to whom.

Systems that implement technical safeguards are usually more costly to implement than systems without an appropriate design and additional protection measures. But it is wrong to state that privacy protection is costly in general. For instance, appropriate technical safeguards against unsolicited mail not only would protect privacy of people but also safe a vast amount of time and thus money to cope with the problem. Companies that pay attention to data security and respect consumer privacy can become much more attractive to consumers than their competitors if the companies communicate the advantage to their customers and the press. This shows that there can be economic incentives for privacy protection that outweigh the cost for implementing an appropriate level of protection. Such incentives should be made use of, see e.g. [Acq02].

2.3.8 Design Guidelines for Technical Regulation

In this section, some guidelines for implementing technical safeguards for privacy protection are derived. In figure 2.8, which is based on one shown in [Sol06], the usual flow of data is shown.

On the left side of the figure, an individual is depicted. Data regarding this individual is collected and stored in databases. The data in these databases can now be aggregated and processed to produce new data that is better usable for the intended purposes. Data that was once collected for a particular purpose can now be used for another. The data can be disseminated to other parties. These parties can now use the data for performing actions that include privacy invasions. The data can also be stored in databases again for anew aggregation and dissemination. The possibility of such iterations is depicted in figure 2.8, too.

CONTROLLING DATA: It makes sense to interrupt the described sequence as early as possible for keeping data under control. Data collection should be limited to an

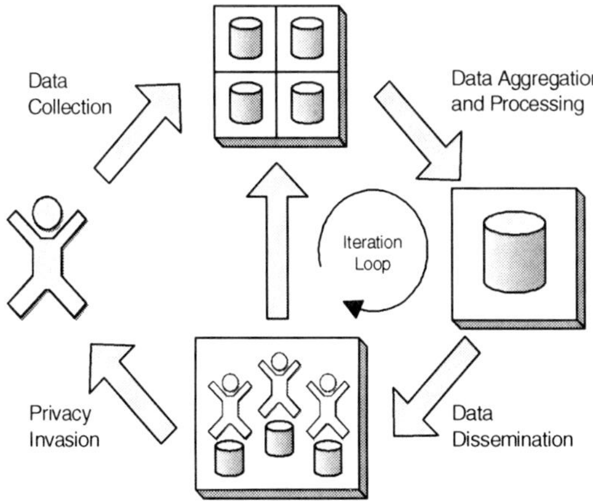

Fig. 2.8. Data flow ending in privacy invasion

amount as low as possible; if possible, anonymity and pseudonymity should be made use of. Once the data is present, data aggregation and processing can hardly be inhibited due to the immaterialness of data (see [Roß05] and [KC04]). Data dissemination should be controlled whenever possible because once the data has been disseminated to other parties, there is no real control over the data present any more to prevent privacy invasions effectively.

In concrete, there should be a means for controlling who gets and has which data to what extent. The individual should know what data is collected and have influence on that process. Langheinrich speaks about "Notice" (to the subject that is being monitored), "Choice", and "Consent" [Lan01] in this context. There should be at least an audit trail what has happened [Sat03] so that the individual is able to react to an unwanted data collection.

After data collection, the individual should be able to see the collected data and be able to perform corrections. Therefore, data should be stored in reach of the individual (see [Lan01]). In analogy to single-sign-on, where authentication is only performed at one place instead of several ones, it makes sense to have only few databases in which data about an individual is stored instead of having data being scattered around.

DATA SECURITY: Data security and privacy are related to each other. Privacy cannot be protected if the system is inherently insecure. This means that the systems need to be designed in a secure manner if privacy is a concern: "Adequate security safeguards should be put in place, according to the sensitivity of the data collected" [Lan01].

SUBJECTIVE DEMAND: As already explained, the value of privacy differs among different people. This results in a different demand for privacy. The consequence is that technical systems need to be flexible to cope with different levels of demand. This is especially important if there is no suitable way for individuals to opt-out, i.e. to decide not to use the system.

TRANSPARENCY: Technical safeguards should be comprehensible. Only this way, individuals can trust and accept technical implementations without worrying about privacy. This does not mean that everybody needs to be able to understand the technical details underlying a system. But the workflows should be that transparent that privacy obviously seems to be protected. Technically skilled users should be able to understand in quite a detail why privacy is adequately protected.

INFORMATION BALANCE: Information asymmetry leads to privacy violations as stated in subsection 2.3.6. One should thus bring information into an appropriate balance. As data collection is reality today, one should rather have a controlled, open access than a hidden access by single parties or government agencies [Bri99].

The goal of keeping information in balance has already been stated in the context of ubiquitous computing in [JHL02] where "The Principle of minimum asymmetry" is proclaimed: "The presence of asymmetric information and negative externalities are at the heart of the information privacy problem.", "Our position is that the role of any technical approach in addressing privacy concerns should be to minimize the asymmetry between data owners on one side, and data collectors and data users on the other."

MANDATORY SAFEGUARDS: The technical safeguards should be introduced in the design phase of a system and become an inevitable part of that system. In other words, technical safeguards should not be easily removable. Changing a system from legal operation to illegal operation or from privacy-enabled to privacy-disabled should ideally be difficult and require the coordination of different parties. The reason is that it should become difficult for a totalitarian government to use the technical infrastructures introduced by a democratic government. This way, people's fear of a surveillance society can be lowered. Rossnagel suggests creating a "Civil Information Society" in which using technical systems for surveillance is at least controllable [Roß02] to prevent a development into the direction of a surveillance society.

CONVENIENCE: Privacy respecting technical systems should be as conveniently usable as possible. A severe problem is stimulus satiation: Humans can only process a limited amount of information. Thus, pervasive computing aims to keep the interaction with users as low as possible, e.g. by context sensing. On the other hand, as stated above, people should be noticed about data collection and use and be able to react on that. A possible solution would be a "personal agent" that applies user defined policies so that only relevant information is passed to the user and other interaction is performed by the "personal agent" on behalf of the user. This way, an appropriate balance between convenience and notice/control can be established.

Even better and more convenient are solutions that act implicitly. For example, permission of performing actions can be based on proximity and locality. An action that happens within reach of a person is more likely to be a wanted action than one that is performed remotely. Using proximity and locality, spatial and temporal borders (see subsection 2.3.6) can be emulated and used for making systems usable more conveniently.

Ideally, users should be assisted in such a way that no user attention is required for privacy protection. No explicit actions should be required to protect one's data.

Instead, the protection should be a side effect of normal interaction with systems and objects.

COST: Implementing technical safeguards for privacy protection comes at a cost. As already explained, the total costs should be kept low by exploiting economic incentives. Further, the level of privacy a system is able to maintain should not be higher than the users' demand if that created additional cost. The focus should be on a proper privacy protection when the risk of an abuse of data is high.

2.3.9 Privacy Summary

Privacy is a central element within this book and has thus been discussed in some greater detail in the previous subsections. It is important to understand what privacy is and why users demand privacy to be able to model systems that implement technical safeguards for the protection of privacy. Thus, this logical chain has been analyzed. The historical development showed that the perception of privacy develops over time. It is also dependent on social and cultural context as well as on the single individual.

The given overview is based on a variety of publications – historical, legal and technical ones – that have been used to create a proper understanding. The goal was to derive a model of privacy that enables to give implementation guidelines for privacy aware systems. This shall enable the modeling of privacy aware RFID systems following fair information practices in the subsequent chapters.

Within this book, it is not possible to perform the social discourse in what kind of world we want to live in the future and to define hard constraints for privacy protection in technical systems. The systems to be discussed and built in the following chapters are offers that are built as much as possible on the principles discussed. The proposed systems need to withstand discussion in the community and possibly need adoption.

2.4 Cryptographic Primitives

Cryptographic primitives are important tools for building secure systems. Techniques like digital signatures are meantime well known due to their wide use in today's Internet. In this section, important cryptographic primitives that are relevant within this book are presented in brief, and aspects that are important for the following chapters are highlighted.

After giving an overview on symmetric key and public-key cryptography, hash functions and their characteristics are discussed. Afterwards, the problem of random number generation is presented. At last, some considerations and directives that are relevant for the remainder of this book are drawn before the section concludes.

This section cannot provide more than a short introduction. There are much more interesting cryptographic primitives than are presented here like key establishment protocols (e.g. Diffie-Hellmann), zero-knowledge protocols or secure two party computation (see [Cra99]) that can be a good foundation for building secure systems. See [HM05] for an example application of secure two-party computation. A comprehensive introduction into modern cryptography and its application can for instance be found in the "Handbook of Applied Cryptography" [MvOV96].

2.4.1 Symmetric-Key Cryptography

Symmetric-key cryptography is the straightforward form of cryptography: The sender encrypts his plaintext using a key. Everybody in possession of that key is then able to decrypt the ciphertext and regain the plaintext. The key must be distributed to legitimate receivers over a secure channel so that an attacker cannot obtain it. Today, such a key transfer is often done with key establishment protocols or using public-key cryptography. Note that the algorithms used in symmetric-key cryptography are not only used for encryption purposes but can also act as building block for other primitives like pseudorandom number generators or hash functions.

There are two classes of ciphers: "Blockciphers" and "Stream Ciphers". Prominent symmetric-key blockciphers are DES (Data Encryption Standard), AES (Advanced Encryption Standard), FEAL (Fast Data Encipherment Algorithm), IDEA (International Data Encryption Algorithm), SAFER (Secure And Fast Encryption Routine), and RC5 (Rivest Cipher). Block ciphers encipher a complete block of data at a time whereas the size of such a data block is usually 64, 128, 192, or 256 bits and the encryption transformation is fixed. In contrast, stream ciphers work on sequences of individual characters with an encryption transformation that varies over time.

2.4.2 Public-Key Cryptography

Public-key cryptography is also known as asymmetric cryptography. In public-key cryptography there are two keys, a private one and a corresponding public one. It must be computationally infeasible to obtain the private key from the public key. The most famous asymmetric encryption algorithm is RSA (Rivest, Shamir, and Aldeman); it is widely used today.

A message that is encrypted with the public key can only be decrypted with the private key. The main advantage compared to symmetric-key cryptography is that the public key can be made publicly available. This way, key distribution is much easier. It is only important to ensure that the public key is really part of the key pair of the intended receiver.

There are also public-key cryptosystems like ElGamal [Gam85] that permit re-encryption: A given plaintext can have different ciphertexts, and one can change the appearance of a ciphertext without deciphering. For re-encryption it is not required to have the private key; to be in the possession of the public key with which the plaintext has been encrypted is sufficient. With "universal re-encryption" [GJJS04] that is also possible with ElGamal, neither the private key nor the public key is required for re-encryption. The underlying mathematic property that is employed is homomorphism: Certain computations on encrypted messages correspond to other operations on the cleartext messages.

2.4.3 Hash Functions

A hash function is a function that maps input from an arbitrary domain to a finite output range. Thereby the input domain is typically much larger than the output range. In this case, the mapping cannot be injective because then all the domain values need to be mapped onto the limited set of range values. This means that a single range value can have several different preimages. The ability of hash functions to map input that is longer than the output to output of specific length is called "compression" (see chapter "Hash Functions and Data Integrity", pp. 321-383, in [MvOV96]).

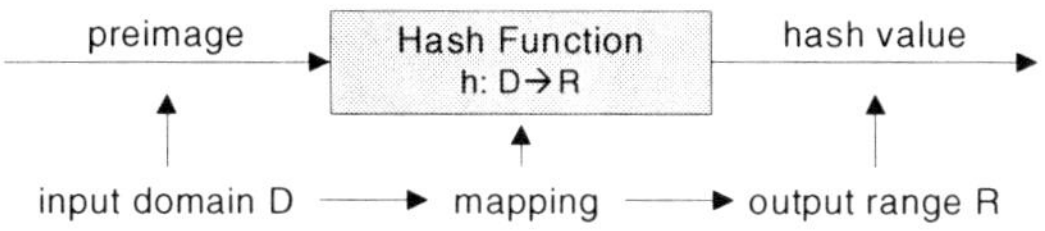

Fig. 2.9. Operation of hash functions

Instances where different preimages yield the same output are called "(hash) collisions". A "good" hash function experiences as few collisions in the domain it has to deal with as mathematically possible. This means that for a domain with 2^t possible inputs and a range with 2^n possible hash values, the number of inputs that are mapped to the same image should be $2^t/2^n = 2^{t-n}$ for each value. Ideally, the probability of an arbitrary input value getting a particular hash value as output should be 2^{-n} so that all possible hash values are equiprobable.

In computer science, hash functions are often used for the creation of hash tables to speed up search for information by mapping complex datasets into handier hash values, i.e. keys (see figure 2.9). The calculation of hash values should computationally be as simple as possible and yield a uniformly distributed output.

In cryptology, hash functions were introduced in the seventies for protecting the authenticity of data. Nowadays they help to solve a variety of security related problems. They are used in various algorithms, e.g. for hash chains [Lam81], which themselves are employed for a variety of purposes. For instance, hash functions are nowadays used to check the integrity of data, for authentication purposes and for storing passwords in a safe manner.

To match this purpose, hash functions for use in cryptology need to have additional characteristics compared to conventional hash functions (see e.g. [MvOV96] for a survey). These characteristics are in colloquial language summed up by the term "one-wayness". The term deals with the reversibility of a hash function as well as with the possibility to find coherences between input and output values of a hash function.

A basic application of hash functions in cryptography is shown in figure 2.10. There the hash function is uses as "modification detection code" (MDC). A hash value of a message is calculated before the message is sent over an insecure channel. If the receiver gets the hash value in a safe manner (usually employing digital signature schemes), he can check whether the message was altered by an attacker or not. The advantage of this scheme is that only few data, i.e. the hash value, needs to be transported via the secure channel. But the example also shows that it needs to be difficult for an attacker to change or construct a message in such a way that it has the same hash value as the valid message.

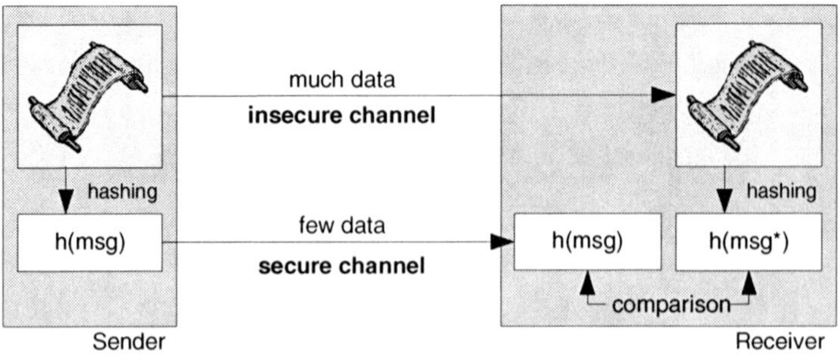

Fig. 2.10. Application of an unkeyed hash function for ensuring message integrity

This leads to the question which properties are required for cryptographic hash functions. This depends on the intended application. The most relevant properties will be explained briefly in the following:

PREIMAGE RESISTANCE: If only an arbitrary hash value is given, it is computationally infeasible to find a corresponding preimage.

2ND-PREIMAGE RESISTANCE: If an input/output pair, i.e. a preimage and a corresponding hash value calculated by a hash function, is given, it is computationally infeasible to find another preimage that hashes to the same hash value.

COLLISION RESISTANCE: It is computationally infeasible to find any collision, i.e. it is "hard" to find any two arbitrary but distinct preimages that have the same hash value.

For example, preimage resistance is required when passwords are stored as hash values. Even if a username/hash value - pair becomes known to an attacker he shall not be able to find a corresponding password. The usually more restricting 2nd-preimage resistance is required for ensuring data integrity like in the example in figure 2.10. Collision resistance practically includes preimage and 2nd-preimage resistance and is thus the most demanding characteristic. But note that examples can be constructed in which that does not hold true (see note 9.20 in chapter "Hash Functions and Data Integrity", pp. 321-383, of [MvOV96]).

Usually preimage resistance und 2nd-preimage resistance are required properties for hash functions that are called "one-way hash functions" (OWHFs), but sometimes alternate terminology is used where the terms "one-way" and "preimage resistant" are used synonymous. Hash functions that satisfy preimage resistance, 2nd-preimage resistance, and collision resistance are usually called "collision resistant hash functions" (CRHFs).

Besides the presented basic properties, depending on the application additional properties might be required. The following three properties are taken from [MvOV96] and can also be relevant for securing RFID protocols. More considerations on such additional properties can for instance be found in [CMR98] where the term "perfect one-wayness" is used.

NON-CORRELATION: Input bit and output bits should not be correlated. Flipping an input bit should flip each output bit with the probability of one half ("strict avalanche criterion").

NEAR-COLLISION RESISTANCE: It should be hard to find two preimages whose hash values differ in only a small number of bits.

PARTIAL-PREIMAGE RESISTANCE: Even if part of the input is known, it should be difficult (i.e. there should be no better strategy than trying by brute-force) to recover the remainder.

It is often anticipated that standard hash-functions meet these properties well enough for practical applications (see also [Wei03]), i.e. that enough confusion (making the relationship between input and output complex) and diffusion (removing statistical dependencies) (see [Sha49]) is created to make attacks difficult.

According to [MvOV96] one can distinguish three broad categories of hash functions: Ones based on block ciphers, ones based on modular arithmetic, and customized hash functions. Hash functions of the first two categories can often reuse system components already existent for other cryptographic primitives whereas customized hash functions can be better tailored to the application thus optimizing performance and complexity. MD4 and MD5 [Riv92] are examples for customized hash functions.

Besides unkeyed hash functions there are also keyed hash functions (see [BSNP95] for definition and applications). Keyed hash functions are also called "message authentication codes" (MACs) if used for authentication purposes. MACs are used to provide data integrity and symmetric data origin authentication. Besides an input of arbitrary length they have a fixed-length second input, the key, that serves as parameter. It must be infeasible for an attacker to resolve this key if input/output-pairs are given ("computation resistance").

Unkeyed hash functions can be extended to become keyed hash functions, for instance, by concatenating the regular input and the key. In particular, if the compression feature of hash functions is used, on needs to be careful because MACs have additional requirements compared to MDCs. At least the "partial-preimage resistance" property that was introduced above is required if an MDC shall be used as MAC. See [MvOV96] for further information. RFC 2104 [KBC97] describes a mechanism for message authentication, called "HMAC", that also uses unkeyed hash functions as MACs.

The presented short overview about cryptographic hash functions should have shown that careful analysis is required to ensure that a selected hash function has the properties that are required within the given protocol. Only this way, the security of the built systems can be ensured. As it is of special relevance within chapter 5, some further considerations about hash collisions will be presented in the following.

Hash collisions

Hash functions should assign the domain elements to the possible hash values in a way that appears random. This means that all possible hash values are uniformly distributed. If a particular hash value is given, the probability that an arbitrarily selected hash preimage yields this hash value as result is $P_{\text{singlecoll}} = \frac{1}{2^r} = 2^{-r}$ in which r refers to a r-bit range of possible hash values. In currently used hash functions, r is 128 or higher, which results in a negligibly low probability that a collision for a given hash value is found. A "good" hash function should have the characteristic that there is no faster method for finding such a collision than checking by brute force the results of different preimages which would require an average of 2^{r-1} trials.

In contrast to the low probability that a collision for a chosen hash value can be found, the probability to find an arbitrary collision, i.e. two preimages yielding the same arbitrary hash value, is much higher. This is due to a mathematical phenomenon that is known as "birthday paradoxon" or "birthday problem": The birthday paradoxon is that if you have group of 23 persons the probability that two of them have their birthday on the same (arbitrary!) day is higher than 50%. Thus, the group of persons required for causing a collision, i.e. two or more persons with birthday on the same day, with a high probability is much smaller than one would intuitively expect. The probability that such a collision appears can be calculated using the formula:

$$P_{\text{coll}} = 1 - \frac{d!}{(d-k)! \cdot d^k}$$

In this formula, d is the number of possible values, e.g. 365 possible days in the birthday problem, and k denotes the size of the group, e.g. the number of persons in the birthday problem. The same formula can be used to calculate the probability that hash collisions occur since the underlying problem is the same. In this case, d denotes the size of the domain of the hash values, e.g. for hash functions with 128 bit values (MD5 for example) $d = 2^{128}$ holds. k is the number of hash values of different preimages that are calculated. The resulting collision probability is then the probability that two or more of the hash values are the same.

For such high values like $d = 2^{128}$ or even $d = 2^{160}$ or more, the formula for calculating the probability of a collision becomes difficult to handle. But for many applications, an approximation of the exact value is sufficient. Sayrafiezadeh presented the following approximations in [Say94]:

$$P_{\text{coll}} \approx 1 - e^{-k(k-1)/2d} \approx 1 - \left(1 - \frac{k}{2d}\right)^{k-1} \quad \text{for } k \leq d$$

where the latter is always smaller than the exact probability and has an error

$$\varepsilon < \frac{k^3}{6(d-k+1)^2}$$

The approximation formula of Sayrafiezadeh is well suited for our needs and will thus be used later on.

2.4.4 Random Number Generation

Many security protocols make use of random numbers, e.g. for the creation of challenges or primes. There are different methods for the generation of such random numbers. One can distinguish between methods that yield truly random bits/numbers and methods that generate pseudorandom bits or pseudorandom numbers, respectively.

For the security protocols, it is important that the numbers are random in such a sense that an attacker cannot gain advantage from knowing the way the numbers were created or that he can exploit statistical weaknesses that enable him to optimize his search strategy to make attacks significantly better than brute force.

Random bits or random numbers can be created using physical sources of randomness (e.g. thermal noise of semiconductors) or based on software using unpredictable values that can for instance be derived from user activity (mouse movement, time between keystrokes, etc.) or system load (CPU usage, network usage, etc.).

Pseudorandom bits or pseudorandom numbers are created using algorithms that create sequences of values that appear to be random but are based on a usually much shorter, fixed "seed". Ideally, an attacker is not able to distinguish between a truly random sequence and a generated pseudorandom sequence.

There are methods for generating pseudorandom numbers that are not proven to be cryptographically secure but that "appear sufficient for most applications" ([MvOV96], p. 173). One possibility that is shown in [Sha83] is to use one-way hash functions, in the following denoted by h, and to use the output or parts of the output of the sequence $h(s), h(s+1), h(s+2), \ldots, h(s+i)$ in which s is a seed that cannot be guessed by an attacker. This illustrates that pseudorandom numbers can be created with the same cryptographic primitives that are used for other purposes.

2.4.5 Implementation Considerations

In this subsection, the use of the presented cryptographic primitives in practical applications and occurring problems will be discussed. Some design decisions that are relevant for the remainder of this book will be based on these considerations.

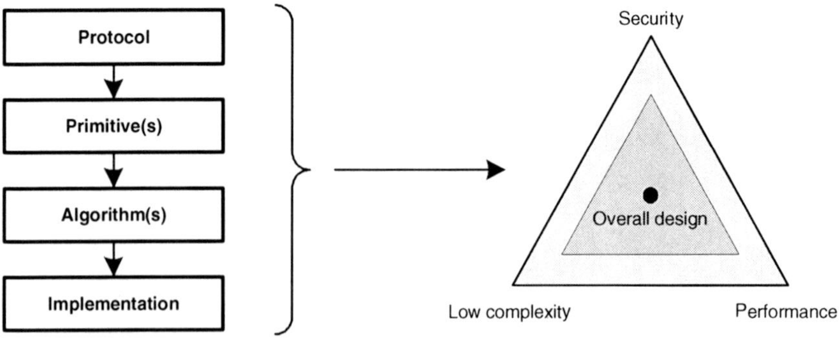

Fig. 2.11. The characteristics of a protocol depend on several layers

As shown in figure 2.11, a protocol uses one or a number of cryptographic primitives. For each of these primitives, an algorithm in a certain configuration is chosen (e.g. RSA with 1024 bit for public-key encryption). The algorithm is then implemented in software or in hardware.

Depending on all these layers, the complete protocol has different characteristics regarding security, complexity, and performance. These properties are at odds to each other, but a good overall design can improve the protocol entirely. Low complexity leads to low costs so that this property is very relevant with regard to economics.

Design and invocation

A very important design step is selecting the cryptographic primitives that shall be used on an RFID tag. Appropriate algorithms need to be chosen for these cryptographic primitives. Afterwards an efficient implementation for the algorithms needs to be selected. All these selection steps have severe impact on performance, complexity, and security of the overall solution. This will be explained in the following paragraphs.

But the overall quality of the solution does not only depend on the proper selection of primitives, the used algorithms and their implementations. The way the primitives are used by the upper layer protocols is also relevant and affects performance and security.

Performance is directly affected by the number of invocations of the primitives within the used protocol and by the possibility to perform different calculations in parallel on the tag to save time. Furthermore, the way the primitives are used affects security. For instance, a protocol might rely on a collision resistant hash function whereas another protocol only relies on preimage resistance to ensure security. This leads to the design principle that upper layers employing the cryptographic primitives as mechanism should be designed in such a way that the amount of data revealed to an attacker is minimized and that the protocols rely on as few characteristics of the security primitives as possible. For example, not relying on the collision resistance property of hash functions increases security or enables to use hash values with half size when the security level should remain the same.

Performance

Public-key algorithms are typically substantially slower than symmetric-key ones. Because of this, these two classes of algorithms are often used in combination: Public-key cryptography is used for small amounts of data and to securely transfer the keys for symmetric encryption. Large amounts of data to be transferred can then be protected using symmetric cryptography. Within the symmetric-key algorithms, block ciphers are generally slower than stream ciphers, at least if implemented in hardware (see [MvOV96]). All these symmetric-key algorithms are generally slower than hash functions. See e.g. [PRB98a] or [PRB98b] for a performance analysis.

This leads to the rule of thumb that hash functions are faster than stream ciphers, stream ciphers are faster than block ciphers, and block ciphers are faster than public-key algorithms. The consequence is that hash functions are the preferred cryptographic primitive if speed counts and the needed functionality can be provided with them.

As already mentioned, the selection of algorithms and their implementation also affects performance. For example, RSA is a widespread public-key algorithm but public-key algorithms based on elliptic curve cryptography (ECC; see [Mil86]) appear to be more appropriate for use in resource scarce embedded devices. The required key sizes for a comparable level of security are much lower in ECC than in RSA; for instance, an 160 bit prime in ECC offers the level of security of an 1024 bit RSA modulus [KMV00], and lower key sizes reduce the required calculation and the required amount of memory and circuitry. This increases performance and also decreases complexity.

Complexity

RFID tags are resource scarce environments. As the number of gates is very limited and speed of reading is an issue, the needed cryptographic primitives should be implemented in hardware. The complexity of all required implementations of cryptographic primitives should be as low as possible to keep the number of required gates and therewith the cost per tag low.

Different algorithms have different requirements regarding the number of gates needed for implementing the functionality, the memory requirements, and the number of clock cycles required for performing the operation. Often, the required number of clock cycles can be decreased if more operations are performed in parallel, i.e. a more complex hardware circuitry is used, and vice versa. The trade-off to be found depends on the application requirements. If the performance requirements are not too tough, one will reduce the number of required gates as much as possible to make the tags as cheap as possible. The latter approach is suggested in [Wei03].

As already stated in the previous paragraphs, not only the selection of cryptographic primitives but also a well thought-out selection of algorithms and their implementations is crucial to keep the complexity low. For example, ECC-based public-key cryptography requires much fewer gates for implementation than RSA-based public-key cryptography. An implementation of ECC-256 is possible with less than 10000 gates whereas about 50000 gates are required for RSA (both in a complexity optimized version) [Van04]. There are many proposals for non-standard algorithms that claim to be efficient but also secure (e.g. very efficient hash functions [KH06] or pseudorandom number generators [JY06]). Such algorithms appear interesting but should only be used after careful analysis by the research community. An overview of the number of gates needed for the implementation of different algorithms can be found in [BMBF07]. As stated there, for SHA-1 [NIST02] about 4200 gates are required; low-cost hashing is stated with 1700 gates.

Security

A proper protocol design on upper layers also enables selection of primitives that have an inherently high security. There are primitives like the "one-time pad" that are unconditionally secure and not only computationally secure. This means that the security of such primitives can be proven information theoretically and does not only depend on complexity theoretic limits i.e. that just an enormous amount of computation would be required to break them.

A problem with all the algorithms whose security relies on complexity theory is that the computational capabilities of attackers have increased enormously over the past decades and – even worse – that regularly new attacks compromising security are found and published. This means that algorithms that are regarded secure today cannot be expected to be secure some years ahead.

For example, the hash algorithm MD4 (Message Digest 4), which was designed by Rivest in 1990, is already obsolete and considered broken, and its successor MD5 (Message Digest 5, see [Riv92]) should not be used as CRHF any more. In 2005, an attack against SHA-1 (Secure Hash Algorithm) was published (see [WYY05]) and is constantly improved so that a transition to other hash functions like SHA-256 or even SHA-512 is starting. The latter have a higher complexity but have an even better performance compared to SHA-1 when implemented in hardware [GLG+02].

To cope with this development, one should provide the possibility of migrating to new algorithms in a way that is transparent for the upper layers. Unfortunately, sometimes this is not possible. RFID tags will be affixed to lots of items, and the cryptographic algorithms will be implemented in hardware. A transition to new algorithms would require changing all the RFID tags in the wild which will not be feasible any more when the tags are used ubiquitously. Even a step-by-step transition is extremely difficult because the life cycle of some items lies in the range of decades. For example, longevity is an issue that definitively should be considered for books in a library [Lin03].

The only viable solution here is to design the protocols in such a way that the impact of vulnerabilities is kept low. For instance, the protocol might be designed in a way so that it is still usable for item identification if the security primitive is broken but that some features like the protection against counterfeiting or the protection of privacy are lost.

To the author's knowledge, long-term security is an aspect that has not been considered in publications regarding RFID security before. Nevertheless it is a problem that necessarily needs attention.

2.4.6 Cryptographic Primitives Summary

In this section, cryptographic primitives that are relevant within the following chapters have been introduced. It has been shown that there is a layering (see 2.11): Protocols use cryptographic primitives that are built using certain algorithms that are implemented in a certain manner. Within the rest of this book, only the upper two layers will be considered any more as everything else reaches too far into the field of mathematics and electrical engineering. The cryptographic primitives are thus seen as black-boxes with certain characteristics without considering special algorithms or even implementation issues.

One-way hash functions appeared to be the most promising primitive for implementation in RFID tags: They are not only powerful building blocks but can also be implemented efficiently compared to other cryptographic primitives. Further, they are a possible building block for pseudorandom number generation.

If not stated otherwise explicitly, within the remainder of this book, the term "hash function" always denotes a one-way hash function that has the required properties. All other kinds of hash functions are not relevant here. Cryptographic hash functions and all other primitives will be used as a black box. It is anticipated that the black box has all properties that are relevant to the specific application.

The selection of the algorithm to be used to fulfill these anticipations and the concrete implementation of the algorithms is not part of this book. Nevertheless, the primitives are used in a conservative way, if possible, to minimize the possibilities for attacks. A guideline will be to limit the number of trials a system is subjected to over its lifetime. This can be done by inhibiting chosen-text attacks or even adaptive chosen-text attacks (see p. 326 in [MvOV96]). Additionally, the consequences of a single successful attack should be kept as low as possible. For instance, a system could be designed in such a way that only a single RFID tag is affected or that the protection of privacy is lost if a hash function is broken but that the system is still available and provides service.

2.5 Summary

In this chapter, fundamentals that are required in the remainder of this book have been introduced. The considerations started with an introduction into RFID technology. Relevant topics like the system components and the layering in communications have been explained. The second section has given an overview of relevant security aspects, and some design guidelines that are relevant for RFID systems have been derived. The subsequent section has dealt with privacy. As privacy depends on user perception and user attitude, these aspects have been examined closely. The result has been design guidelines for design and implementation of privacy aware systems. Finally, important cryptographic primitives have been discussed. One-way hash functions with certain properties have been selected as most promising primitive for implementation in RFID tags. More information regarding the content of the sections of this chapter has been given in section summaries at the end of each section.

3

Analysis and Modeling

In recent years, provoking headlines like the following appeared in press: "RFID threatens privacy", "RFID tags: Big Brother in small packages", and "Are you wearing track shoes?". Even headlines like "Cradle-to-Grave Surveillance" could be read. There are even people that interpret the following excerpt from the Bible as related to RFID:

"16 And he causeth all, both small and great, rich and poor, free and bond, to receive a mark in their right hand, or in their foreheads:
17 And that no man might buy or sell, save he that had the mark, or the name of the beast, or the number of his name."
Revelation 13: 16-17

This shows that security and privacy in RFID systems is a relevant topic that deserves further consideration. On the one hand, in media problems that are not based on the RFID technology itself are presented within the scope of RFID. Thus, many contributions do not hit the center of the problems and sometimes even present them in an overstated manner. On the other hand, the RFID technology has the potential to change our world completely – like electricity or the automobile did in the past. It can bring more convenience and greater productivity but can also harm our privacy and can even become a means for total surveillance.

The risks associated to the widespread use of RFID technology cause justified reluctance and low acceptance. Increasing convenience and productivity are desirable goals, but heading towards these goals while ignoring the unintended consequences is not the right way to go.

In the following section, some examples of security and privacy violations are given as a motivation. Afterwards, the threats are listed and discussed in section 3.2. Based on these threats, the goals that RFID systems should fulfill are described in sections 3.3. A variety of challenges need to be addressed. These are considered in section 3.4). The proximate section 3.5 categorizes different attackers according to their capabilities which can be of various kinds. Depending on these capabilities, attacks with different perceivable impact are possible (section 3.6). As a reference point, the current situation regarding the Electronic Product Code and its backend systems is presented in section 3.7. The chapter closes with an assessment of overall RFID security and privacy (section 3.8) and with a chapter summary.

3.1 Motivating Examples

In the following, some examples are given that demonstrate that security and privacy is at risk when using current RFID technology. There are various threats so that not all possible scenarios are covered by the examples. Furthermore, a counterfeiting problem is presented that can be addressed using RFID technology.

Product tagging of tire manufacturers

RFID tags are used to tag products with the intention to simplify stock-keeping. If these tags are not removed or even more sophisticated devices are used, e.g. to detect the tire pressure, the location privacy of the car holder is at risk. For instance, if chained gas stations install readers at petrol pumps, they can create profiles who is buying gas on which gas stations and how often. They could also reconstruct the travel route of the car. Such readers could be installed by other parties to recognize cars, too.

A potential linking of the identifiers of RFID tags affixed to tires with the identity of the car owners using credit cards or the like even enables the creation of personalized profiles instead of anonymous profiles only.

Product tagging of clothing manufacturers

Whereas the tracking opportunities in the previous paragraph regards cars, with product tagging of clothing, people can be profiled, recognized, and tracked.

Of course, tagging of clothes has interesting applications, e.g. for stock-keeping, at checkout, or in case of replacement. There are also “intelligent home” applications envisioned: A washing machine could warn if there are clothes in the washing machine that cannot be treated together or in case a washing program is selected that is not appropriate.

On the other hand, there are also many negative scenarios. Just when passing by with an RFID reader, one can gather which brand the clothing has if the data is present on the tags. Such data can be used to differentiate between financially strong customers and ones with less deep pocket. It also gives information about consumer buying habits.

If the same tag identifier and thus the same piece of clothing is detected multiple times by the same reader, one can derive that with a high probability all the times the same person is present. If there are many networked readers, for instance operated by a supermarket chain, it is even possible to track and record the movements of that person. Repeated reading of the same tag at the same location in addition reveals information about a person's habits: For instance, that somebody wears the same underpants several days in sequence.

The following citation shows that even much more severe privacy violations are possible even with current technology. The emerging possibilities are feared by some people: "Without any regulation, for example, law enforcement could use RFIDs to monitor people's behavior. Police now routinely videotape public protests; in the future, they'll be able to walk around with RFID readers and collect the serial numbers from people's clothing and other tagged items they're carrying. Matching those serial numbers with retailers' records would yield a list of protesters' names, addresses, and so on. Or police could just look for the serial numbers themselves, at an airport security checkpoint, say. 'That tube of strawberry Chapstick [i.e. a well known lipstick manufacturer; note of the author] was at the World Bank protest! Pull that passenger aside!' Though that level of surveillance may be way down the road, says Albrecht, its implications are unsettling" [KC04].

Electronic passports

Within many countries, new passports contain an RFID tag. It contains in encrypted form the data that is written in clear text on the passport. A digitized picture of the passport holder and in future also biometrical data like finger prints are stored on the tag, too. Many researchers like Schneier question the security of current electronic passports (see [Sch06e]).

The digitized picture is encrypted using a key that is based on data written in clear on the passport. Thus, anybody that is able to look at the passport to gather that written clear text is also able to get a digital photo of the passport holder. As one is able to take a picture of a person by other means, this poses not a high additional risk, but if fingerprints and other biometrical data could be read unattendedly by the passport holder in the future, this could cause more serious problems.

Another problem with current electronic passports is that the data stored on the RFID tag can be copied easily. There is no special protection at all. One could argue that, albeit the data can be copied, it is not possible for a passport forger to alter the stored photo to match another person to prevent a forgery of being detected when the face of the passport holder is compared with the stored photo by an official. That is true, but valid data on an RFID tag is no longer a proof of the authenticity of a passport. A copy of data of the RFID tag can also be used to pass passport control when the photo is not checked. This could be used to enter the security area of an airport using data of another person's passport.

Further, the encrypted data on a passport is a means for unique identification of a person which enables to track the movements of a person. In press, even the threat that terrorists could create bombs triggered by RFID data is stated. These bombs detonate when a certain ID document and therewith the person to be attacked is near. Shielded passport covers can be used to prevent reading data on the tag when the passport is not open.

Product counterfeiting of pharmaceuticals

Product counterfeiting is a huge problem today. According to [ICC04], the value of counterfeited and pirated goods has been estimated at over 500 billion euros annually for 2004 with a rapid escalation expected. But counterfeiting is not only a monetary problem: In the case of counterfeiting of medicines, it can even lead to death of people [WHO06]. Consequently, the drug market is bound to tight regulation in many countries, e.g. by the FDA[1], and technical countermeasures against counterfeiting are welcome.

The question is whether RFID technology can only be used for logistics applications or it can also be used as protective measure against counterfeiting. For the latter to work, RFID tags need to prove genuineness of original products to patients and other parties along the supply chain beginning at manufacturers. This can be done by preventing RFID tags to be cloned or at least by becoming able to detect that tags have been cloned.

3.2 Threats

The examples in the previous section showed that a variety of threats regarding security and privacy appear when RFID systems are operated. In addition, RFID shall be used to increase security so that threats for breaking the new security measures appear. As shown in section 2.1, an RFID system consists of tags, readers, and the backend. All these three entities and also the communication paths between them can be the target of an attacker.

Within the backend and for the communication network between readers and backend systems, IT security can be provided just like in other networked systems. For instance, data is stored in databases, and with a proper access control set-up only legitimate users and services can access that data. Data can be enciphered on the network links using standard schemes like Secure Sockets Layer (SSL) or Transport Layer Security (TLS) [DA99]. As there is no practical difference compared to other networked systems, there is no special consideration required here.

Readers perform the communication with the RFID tags over the insecure wireless air interface and communicate with the backend over a network link like explained above. Readers usually do not store data over a longer period of time so that there is no special threat. But it should be ensured that an attacker cannot get control over readers that they do not own. As readers can also be operated by an attacker, they should be regarded as untrusted devices.

The most interesting threats are those regarding the tags and the ones resulting from the communications channel between tags and readers. Because of that, these threats will be considered in more detail in the following. Note that many of these threats are already present in traditional barcode systems in similar form. But in

[1] FDA – U.S. Food and Drug Administration

RFID systems, a line of sight is no longer required, which makes the threats appear on a broader scale.

ILLEGITIMATE READING OF DATA: If data is stored on RFID tags, an attacker might have an interest in reading out this data. For instance, an attacker could read out tags and with the data on them identify items that are worth of being stolen. There is a variety of possibilities how an attacker could achieve such an illegitimate reading. This will be discussed later on in this chapter. It helps not to store data on tags but in the backend instead because the data can be secured much better there. Especially for confidential data that does not need to be stored in smartcards or RFID tags distributedly, it thus makes sense to store it in backend systems. This has the positive side effect that less memory for data storage is required on the tags. This makes tags less expensive.

Currently, there a no safeguards implemented in standard RFID tags that prevent illegitimate reading of data. As the examples in the previous subsection have shown, even few data stored on an RFID tag like manufacturer or product type can lead to privacy violations. If additional data is stored on tags like in the electronic passport scenario, the problems become even more severe.

Even if there are access control mechanisms implemented on the tags and if data is encrypted outside the tags, data on RFID tags cannot be regarded safe. The data can be extracted by physical means bypassing the normal protocols. Physical attacks that are known from smartcard security can be applied by an attacker. For example, side-channel attacks allow conclusion regarding the inner state of tags and can thus reveal stored data. Implementing protective measures is usually too costly for RFID tags. Because of that, no confidential data should be stored on tags, especially if there is a high incentive for an attacker to get that data. For example, storing cryptographic keys on a tag is not a good idea – at least if the keys are shared amongst many devices.

EAVESDROPPING OF DATA: Eavesdropping data on the communications link between tags and readers results in the same risks regarding security and privacy as illegitimate reading, which was discussed in the previous paragraph.

The problem is that the communications channel between tags and readers is public and shared. Everybody who is near enough can eavesdrop the communication and get the data that is sent and received. As passive tags are powered by the reader, the forward channel from reader to tag has a stronger electromagnetic field than in the opposite direction. This makes the forward channel be eavesdropped more easily than the backward one.

CLONING OR MIMICKING OF TAGS: Within many application scenarios, RFID tags are used to uniquely identify items and to ensure authenticity of those items. This requires that tags cannot be copied easily. A tag that is tightly affixed to the object and that cannot be cloned easily can be used to proof genuineness of the object and can thus be used to prevent forgery of the object.

For instance, transponders are used for immobilizer systems in cars to ensure that the keys are not copied illegitimately. Other applications in which cloning of tags must be prevented are, for instance, access control systems and the prevention of forgery of banknotes or pharmaceuticals.

Prevention of cloning and mimicking is important for high priced products whose removal from a shelf shall be detected immediately: If an attacker could place a mimicking device or a copy of the tag instead of an object with a genuine tag, then the object could be removed unnoticed. In contrast, if an item has an RFID tag that cannot be imitated using a mimicking device, a robber cannot steal the item easily.

In this context, the term "cloning" designates the process of creating an exact logical copy that is not distinguishable from the original tag on protocol level. This means that for any reading device the original tag and its copy behave the same and are thus indistinguishable from each other: This way, both appear to be genuine.

Even if tags are indistinguishable on protocol level, the original and its copy are distinguishable below protocol level, i.e. on the physical layer: For instance, even if devices are identical in construction and in memory content, there are minor differences due to tolerances of electronics that lead to differences in power consumption or time response that can be detected with appropriate equipment but not using ordinary readers. The same holds true for the optical appearance of the original and its copy: For example, a tag in a blue package cannot be distinguished from a one in a red package by an ordinary reader. Such other possibilities for ensuring authenticity are not considered within this book.

For successful temporary mimicking of tags without special protection, an exact copy of a tag or its functionality is usually not required. The reason is that most operations concerning a tag are simple reads of its identifier or data; other operations occur comparatively seldom. Thus, a mimicking device that supports the read operations is often sufficient for an attacker: For mimicking it is sufficient that the original tag appears to be in place, at least for some time. Thus, only for perfect mimicking that will not be detected anytime, a fully functional copy of the original tag is needed by an attacker.

A cloned tag is in principle capable of mimicking an original tag perfectly. But if tags maintain state-information, the clone and the original can get different on further use. This happens when state-information in the used tag changes due to protocol operations. The state-information in the other tag might afterwards not meet the expectations of the reader or backend any more so that the tag is regarded invalid. This limits the possibilities of an attacker: Imagine he manages to clone a tag. Unlike in the scenario of a door lock, where a copy of a key can be used anytime later on, the cloned tag might become useless if the original one is still in use. The attack has failed in this case. Similarly, the use of a cloned tag might render the original one unusable.

Mimicking of tags can in addition have the following implications: It can be used for infiltrating incorrect data into an RFID system. In a scenario with networked readers, an attacker could pretend that an object is at a location in which it physically is not. This can have severe security implications if the location is used as a means for authentication. Furthermore, logistics applications can get confused if items appear at locations that are not expected. Unfortunately, such consequences of false

RFID data are not yet investigated in RFID literature and actual implementations in a depth that would be adequate to the possible negative implications. In [BFHF03], at least temporary failures of reading tags caused by environmental conditions are considered and solved by multiple reads and data aggregation.

RECOGNITION OF OBJECTS: RFID tags can be used to detect that a person or an item is at the same reader as a time before. This possibility results from the core functionality of an RFID system: to identify objects that are queried by a reader. This functionality is essential but can on the other hand result in privacy violations. Recognition of persons can be used to explore customer habits.

Persons can be recognized using RFID tags that they carry, for instance, implanted under the skin. Such devices are already in use today for access control to buildings and keeping medical information. Another possibility is to recognize persons by the objects they carry regularly. These objects need to have an RFID tag affixed. Examples are wristwatches, glasses, or shoes. If there are items that are not carried regularly by a person but with a certain probability, detecting several of these items at the same time results in a high probability that the person in question is present. This is a form of tracking by constellation.

Currently, recognition of persons is not a significant problem because there are not many RFID tags affixed to items that people use regularly. Exceptions are RFID chips in library books and staff ID cards. If RFID tags become ubiquitous, for instance, in ID cards, credit cards, customer cards, banknotes, clothing, etc., recognition might become huge a problem. The possibilities of abuse are discussed in press. There is no mutual consent on the extent of the problem. However, there is a fear that unwanted recognition could result in the creation of more detailed customer profiles and unwanted surveillance.

TRACKING OF OBJECTS: Passive RFID tags have a rather short read range. Nevertheless, if readers are networked and operated by the same organization, this organization can use the infrastructure for tracking purposes. Tracking of objects is relevant for many applications. Currently, tracking is widely used in logistics: Parcels and letters are read by many RFID readers along the delivery path so that the next steps can efficiently be planned and customers can get an idea where their shipping goods currently are.

Whereas tracking of items is often part of the intended application, tracking of persons is often not desired and can result in severe privacy violations. Such a tracking of persons could occur for example by using personalized tickets for public transportation. Even if persons do not directly carry personalized RFID tags, the tracking can be performed by querying objects equipped with RFID tags that the persons carry. Again, items like wristwatches or glasses enable a rather direct mapping, whereas other items only allow tracking by detecting item constellations. The threats are unwanted creation of movement profiles and an abuse of the infrastructure for surveillance by a totalitarian government.

Currently, unwanted tracking is not a severe privacy problem. RFID tags are usually used in closed-loop applications. There are usually no networked readers that share data amongst several companies or organizations, and if so, these readers are out of range of “ordinary” people. But when readers become ubiquitous due to falling costs and the variety of RFID applications, one can expect a trend to network the readers to get more fine-grained data. With a dense network of networked readers, the infrastructure is present for an efficient tracking of people. Such a tracking is already possible today using mobile phones. But here people are able to opt-out, i.e. to decide to switch off their mobile phone. In contrast, people will not be able to escape a world in which RFID tags are ubiquitous.

CAUSING MALFUNCTION: Attackers might have an interest in rendering an RFID system malfunctioning. For instance, in a checkout-less store that is envisioned by some supermarket chains, customers might be interested in their goods not being detected so that they need not pay for them.

It is not possible to secure RFID against all kinds of attacks. For example, physical destruction or chemical treatment of tags cannot be prevented effectively. Fortunately, currently there are only few areas of application in which attackers might have an interest in causing malfunction. But the possibility of such kinds of attacks must be considered. If possible, attackers should get no incentives to perform such attacks. If this is not possible, there should be means implemented to detect and react to such attacks.

3.3 Goals

Considering the threats that have been presented in the previous section, one can derive the goals that security and privacy respecting RFID systems should reach. The result is shown in figure 3.1 from an abstract point of view. The goals correspond roughly to the threats presented in the previous section but cannot be mapped one-to-one.

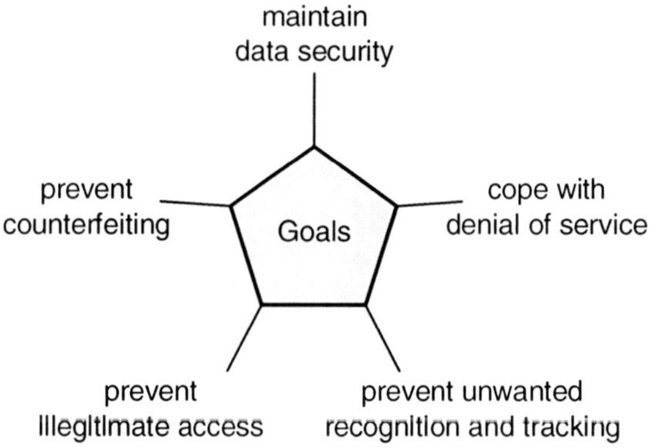

Fig. 3.1. Goals for security and privacy respecting RFID systems

There are five goals: maintaining data security, preventing counterfeiting, preventing illegitimate access, preventing unwanted recognition and tracking, and coping with denial of service. Ideally, all these goals are not only reached in closed systems but also in systems that aim at sharing of RFID data amongst multiple organizations.

MAINTAIN DATA SECURITY: Illegitimate reading of data must be prevented in RFID systems, because the data must be treated confidentially since it may be privacy sensitive. A "good" RFID system must be able to cope with the threats regarding illegitimate reading of data that have been presented in section 3.2.

PREVENT COUNTERFEITING: For many applications, preventing counterfeiting is a goal that should outweigh the higher costs of RFID compared to optical barcodes, which can be copied easily. If RFID tags only emit unique numbers for identification, they can be copied or mimicked easily. But with RFID tags that can prove their authenticity, counterfeiting can be prevented much more effectively.

PREVENT ILLEGITIMATE ACCESS: Preventing illegitimate access is related to maintaining data security. For the latter, access to stored data – whether stored on tags or in the backend – needs to be controlled. But preventing illegitimate access is also a prerequisite for ensuring the integrity of the data in an RFID system. Illegitimate access to system components enables the infiltration of false data.

PREVENT UNWANTED RECOGNITION AND TRACKING: Recognition and tracking of objects are core functionalities of RFID systems. For privacy reasons, if persons get involved, this functionality is often no longer a wanted one. This is a severe conflict that needs to be solved: Sometimes the functionality is wanted and sometimes it is not, and there need to be technical models that provide a suitable trade-off.

COPE WITH DENIAL-OF-SERVICE: This goal is directly connected to the availability of RFID systems. Even if attackers try to put a system out of service, ideally the system should keep running and provide service to legitimate users. A prerequisite is that the integrity of the system is preserved. As it is not possible to prevent all kinds of denial-of-service attacks, RFID systems should at least provide means to cope with denial-of-service attacks, e.g. by implementing means for detection and recovering.

3.4 Challenges

Reaching the goals presented in the previous section poses a variety of challenges. In this section, an overview will be given.

CONFLICT OF OBJECTIVES: As already stated in section 3.2 on threats, a means for identification and tracking is sometimes a required functionality of RFID systems, but at other times it can violate people's privacy. Different parties, e.g. logistics companies and consumers, have different needs. It is a challenge to design RFID systems that easily adapt to these different requirements since the trade-off depends on the intended applications.

MULTIDISCIPLINARITY: Building RFID systems adhering to security and privacy needs to bring together multiple disciplines. Computer science is required for designing communication protocols and middleware. Electrical engineering realizes the required functionality in hardware efficiently and is responsible for the physical layer of communication between tags and readers. Mathematics provides basic cryptographic primitives and theory of probabilities for different areas. Economics dictates constraints imposed by the laws of market and assesses the real world applicability of approaches. Social sciences bring in the requirements of the users by covering topics like privacy and usability. Law gives the legislative basis for the interplay among people and organizations.

REQUIREMENTS: Besides the presented goals regarding security and privacy, there are three additional requirements: low cost, coping with few tag capabilities and resources, and inter-organizational operation. RFID tags are to be manufactured in large quantities because in the future they shall get affixed to every object. Thus, the price per piece is a very important quantity. To keep the price low, the RFID tags cannot be equipped with much resources (processing power, memory) or other features like special coatings. Thus, the features of RFID tags and costs are at odds. This means in practice that one has to cope with as few features as possible to keep costs low. The third requirement that is currently emerging is inter-organizational RFID systems. Data that has been collected by one company shall often be shared with other companies. This way, the network of readers gets denser and one can get more fine-granular location and tracking information, e.g. in logistics applications.

ADDITIONAL REQUIREMENTS: There are a variety of additional parameters that are relevant for "good" RFID systems. A very important aspect is *scalability*. In a world with billions of objects that have an RFID tag attached, scalability of the overall system is crucial. Even in such dimensions, the system shall operate reliably and in the intended manner. Thus, *dependability* is an issue, too. For keeping costs for implementation and the probability of security flaws low and for making the systems well maintainable, the systems' *complexity* should be as low as possible. *Robustness* is an important property, too: Tags are usually passively powered and thus need to cope with sudden loss of power. They must not come into illegal states. For interoperability, there should be *standards* for inter-organizational operation of RFID systems. Such standards need to be established timely so that early adaptors do not implement incompatible technology. For practical application, safeguards regarding security and privacy should not limit *read range and speed of reading* of tags considerably. If cryptographic primitives are used, one needs to consider *migration paths* for the case that the primitives get broken over time. Ideally, the impact of broken primitives should be restricted. For instance, a system would loose its privacy features but still be able to operate. From a user's perspective, *transparency* and *usability* are also relevant requirements. This includes a user's understanding of the main processes taking place and an easy handling resulting in the ability to use the RFID system conveniently. These two aspects have already been explained in the design guidelines for privacy respecting systems in subsection 2.3.8.

3.5 Attacker Capabilities

Section 3.2 discussed the threats that users of RFID systems are exposed to. On the other side, there are attackers with certain capabilities. These attackers try to counteract the goals presented in section 3.3.

In this section, the attackers are categorized according to their capabilities. Although there is a variety of capabilities that can appear in different combinations, it makes sense to simplify the scheme to get a practically usable categorization. This can be done without loss of generality because if an attacker has a certain capability, one can assume which additional capabilities of less power he probably has as well.

In figure 3.2 the proposed categorization is presented. The different classes of attackers are denoted by a name ("dramatis personae") for providing references. This has been done in the style of the scheme presented [Wei03], but the scheme has been substantially revised and extended.

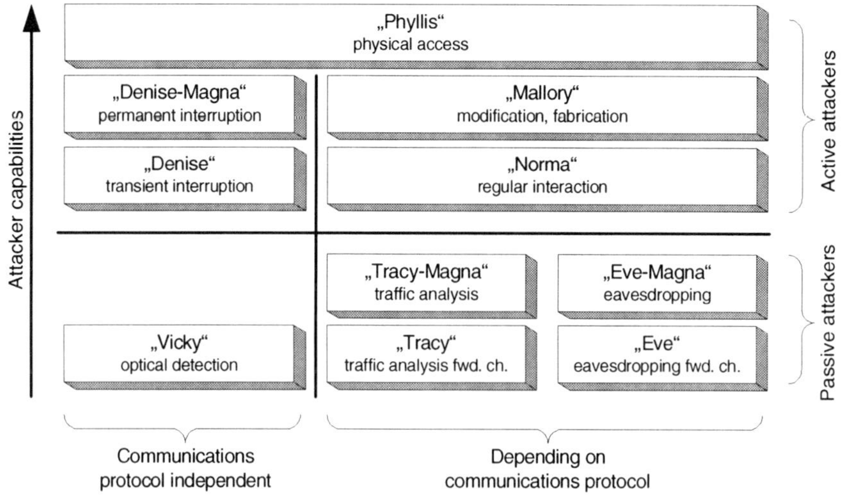

Fig. 3.2. Categorization of attacker capabilities

From bottom to the top of the figure, the capabilities of attackers are increasing. In practice, just like the capabilities also the efforts for attacks increase from the bottom to the top. Thus, one can expect that attackers in the top part of the figure will only operate if the incentive in form of a commercially usable attack result is high enough, whereas attackers on the bottom part need much less incentives since the required efforts are lower.

The attackers are categorized horizontally into ones that work independently of the communications protocol and ones that are dependent on the communications protocol. The term "communications protocol" denotes the way of communication between readers and tags on the physical layer and the data link layer, e.g. protocol messages (see section 2.1.4). This means that on the left side of the figure attackers do not need knowledge about the communications protocol or need not use this knowledge, respectively. In contrast, attackers on the right side of the figure make use of certain properties of the communications protocol, e.g. protocol states or just the amount of data sent.

Vertically in the figure, attackers are categorized into active and passive ones. Passive attackers do not emit electromagnetic fields or waves and are thus only watchers or listeners. Active attackers can emit electromagnetic fields or waves and can thus actively take part in communication processes or perform other actions.

Presentation of attackers

In the following, the capabilities of the attackers shown in figure 3.5 are explained in more detail. The order of presentation corresponds to the attacker capability, i.e. the presentation starts with the weakest attacker and ends with the most powerful one.

VICKY: Vicky is the weakest attacker and could thus be regarded as no attacker at all. Vicky's capabilities are limited to detecting the presence of an RFID transponder passively. Her only means to do this is visual inspection: She can see tags. Vicky can identify victims for attacks performed by more powerful attackers.

TRACY: Tracy can perform traffic analysis of data that is sent from readers to tags. She is a passive listener that is not able to understand the content of the messages sent. This means that she is only able to gather when and how often data is sent. She can also gather the amount of data that is sent but cannot interpret that data. Thus, Tracy can only deduce information from patterns in communication.

It does not matter whether data is encrypted or not for Tracy because she does not understand the data anyway. Nevertheless, traffic analysis on collected data can be useful for attacks. There is a nice anecdote from World War II (found at http://www.freeswan.org/freeswan_trees/freeswan-1.5/doc/glossary.html): By gathering that all German radio traffic stopped, the British could assume that an attack was coming. The "radio silence" order that was intended to preserve security actually gave away the Germans' intention.

In RFID systems, traffic analysis shows the presence of readers and tags. This can be the basis for attacks by more potent attackers but does not affect the goals stated in section 3.3. But the characteristics of data like its amount can enable unwanted recognition and even tracking of persons. For instance, if a person carries an unusual high (or low) number of tags and the same number of tags is read by readers more than once, one can derive that the same person passed the reader. A mix of different tag standards, e.g. tags using different frequencies or different anticollision protocols, can also be characteristic.

Problems with traffic analysis are given everywhere in the Internet when anonymity shall be provided and communication relationships be concealed. Cryptographic protocols like IPSec [KS05] do not give protection so that dummy traffic and mix servers need to be employed to confuse the attacker. The topic will be discussed in more detail in chapter 5.

EVE: Eve has extended capabilities compared to Tracy. Eve can not only perform traffic analysis but can also read and interpret the data sent from readers to tags. She is thus an eavesdropper to the reader-tag link of communications. Eve can counteract the goals presented in section 3.3 at two aspects: She can use her capabilities to perform unwanted recognition and tracking and can also break data security.

Data that remains the same upon different queries of the tag and that is eavesdropped by Eve can be used for recognition and tracking. Whereas Tracy could only use characteristics like the number of tags that a person carries, Eve can use the content of data messages sent from reader to tag to distinguish between different tags. In anticollision schemes (see RFID introduction section 2.1.4) like binary tree walking, the tags are addressed by the reader. If the tag address equals the unique tag identifier, which is usually the case, Eve can uniquely identify tags. This way, recognition of tags and therewith persons becomes very simple for Eve. Encryption of data is no solution here because as long as tags contain constant data – whether encrypted or not – that is addressed by the reader, it can be used as distinguishing element.

Eavesdropping can also break data security. If confidential data is sent in clear, Eve can overhear the data and then abuse it. Encryption can protect data here but is costly to implement in low-cost tags.

TRACY-MAGNA and EVE-MAGNA: Tracy and Eve can only operate on the data sent in the direction reader to tag. In contrast, Tracy-Magna and Eve-Magna have the same capabilities as their correspondents Tracy and Eve but can perform their analysis in both directions of communication.

The reason for the distinction of the two scenarios is that, as already explained in subsection 2.1.4, the field strength on the communications link from reader to tag is much higher than in the opposite direction. This is due to the fact that passive tags obtain the power required for operation from the reader.

The capability to analyze both directions of communications gives Tracy-Magna and Eve-Magna additional possibilities to perform their analysis. For instance, Tracy-Magna can also use the amount of data that is sent to the reader as an additional distinguishing feature. Thus, even if the number of tags is the same, the constellation of the amount of data on the tags can enable recognition of a person. Similarly, Eve-Magna can use any identifying data that is sent from tags to reader as distinguishing features. This way, even if blinded tree-walking is used as anticollision scheme, which does not sent address data in reader-tag direction, a tag can be recognized by Eve-Magna. Additionally, data stored on tags that is sent in clear to the reader can be overheard by Eve-Magna which breaks data security.

DENISE: Denise can interrupt the communication between readers and tags temporarily. Thus, she can temporarily put parts of RFID systems out of service.

There is a variety of possibilities to do so. One way is shielding readers or tags, e.g. by using aluminum foil. Here, Denise acts only passively. One can buy wallets today that include a metal mesh acting as a Faraday cage. Therewith, unwanted reading of tags within the wallet can be effectively prevented.

Another possibility is to disturb the communication between readers and tags by using jamming transmitters. In this case, Denise becomes active by emitting signals. If the communications link is thus too noisy, normal operation of the RFID system is no longer possible. Note that Denise needs not have any knowledge about the communication protocols used.

NORMA: Norma acts as a regular tag or a regular reader within the RFID system. She uses the communications protocol on all layers in the way it is intended in the specification of the protocol. She does not try to cheat or to introduce invalid data or malformed messages.

This means that Norma can query tags just like any legitimate reader: She can use the same request and response messages as such a legitimate reader. Similarly, Norma can act as a regular tag and can thus be queried by a reader like any other legitimate tag.

In the presented point of view, each reader that is not a legitimate reader for a certain tag acts as an attacker with Norma's capabilities: It queries the tag just in the way a legitimate reader does and can obtain the publicly available data from the tag.

Compared to Eve-Magna (and also the other Eve and the two Tracies), Norma does not need to wait until a tag is queried by another reader to obtain information. Norma can query a tag by herself to obtain the publicly available data from it. Depending on the protocol, Eve can overhear passwords or access restricted data that Norma cannot. Norma can only guess passwords and try by brute force until she guessed right to obtain data. But this way, she potentially can obtain data that would not have been sent in standard tag queries.

If Norma gets the data stored on a tag, Norma can use this data to mimic this tag. She does the same what the original tag would do: Take part in the communication process in the regular manner and send its data when requested. Getting the data stored on a tag requires the data to be available to Norma: If the data is publicly available on the original tag, Norma can act as a reader and query the tag for its data. Alternatively, Norma could employ Eve-Magna's eavesdropping capabilities to overhear the required data in a tag query that is performed by another reader. The latter alternative is sometimes the only viable action for Norma: If passwords or the like are required to get the data on a tag, eavesdropping a legitimate read reveals even data that is protected in a simple fashion.

Mimicking of tags is equivalent to counterfeiting: The attacker becomes able to pretend that a tag is there when it no longer is. This way, an attacker could remove a high valued item with an RFID tag affixed by replacing the item (including the original tag) with a cheaper mimicking device. If RFID tags are used for access control systems, e.g. in form of an ID card, counterfeiting is also a severe problem: If it is possible to copy tags, tags no longer provide protection and become a bad security device. Unwanted copying of tags is also a problem when they are used to combat product counterfeiting, e.g. for pharmaceutical items.

Whereas Eve-Magna operates passively and thus virtually undetectably, Norma needs to take part in the communications process actively. If Norma acts as a reader, she has to provide the power for passive tags and needs to perform the queries. If she acts as a tag, she answers when she is queried by a reader. Thus, Norma's operation is detectable in principle.

DENISE-MAGNA: Denise-Magna has just like Denise the goal to disrupt normal operation of the RFID system. But in contrast to Denise, Denise-Magna has the capability to cause permanent disruption of service.

A permanent denial of service can be achieved by destruction. There is a variety of possibilities to render tags unusable: Mechanical or chemical treatment or highly powered electromagnetic waves are effective means to cause permanent service disruption. Very high or very low temperature or other environmental forces can lead to permanent malfunction, too.

In practice, vandalism could be the cause of mechanical demolition of readers. Massive mechanical bending of tags can also render them inoperable. Removal of the antenna of a reader or tag also makes the devices inoperable. There are vendors of RFID tags that use tag antennas with predetermined breaking points. By stripping the upper layer of the RFID label, the tags can be rendered inoperable by customers to protect their privacy.

MALLORY: Mallory is a very powerful and thus dangerous attacker. She can take part actively in the communication process between readers and tags and can perform a variety of actions. There are three different roles of participation conceivable for Mallory: Acting as reader, acting as tag, or working as man-in-the-middle between reader and tag.

Mallory can do whatever she wants with messages that are involved in the communication process: She can create messages, alter messages, and cause message loss. Thus, she has all the capabilities that are required for spoofing attacks, i.e. fooling communication peers within the chain of communication.

She is not bound to legislatory field strength limits. This means that active attacks can be performed from distances larger than the usual nominal read ranges. In [KW05] practical examples are given: Low-frequency (LF) tags that have a read range of about 10 cm can be queried from a distance of about 50 cm. LF tags can be mimicked from a distance of about 50 m by transmitting directly on the sidebands instead of performing inductive load modulation: A reader cannot tell the difference between a regular tag a few centimeters away and such a mimicking device that is several meters away. For instance, an attacker can thus perform attacks by accessing readers without going into the range of installed security cameras.

Mallory can use design flaws in the communication protocol to bring backend, readers, or tags into unwanted states. This way, she can cause transient or even permanent disruption of service like Denise-Magna.

Weaknesses in the communication protocol or in the implementation of hardware and software of tags and readers can also be used to obtain data that would not have been provided within regular operation of the RFID system. This can break data security by revealing confidential data. Obtained data can also be used for unwanted recognition or tracking. If all data that is stored on a tag is revealed by Mallory, an attacker with the capabilities of Norma is sufficient to mimic that tag on protocol level perfectly. The problems related to counterfeiting or introduction of false data into the system that have already been explained in the paragraph about Norma and in the section about threats (section 3.2) then reappear here.

The ability to query tags in arbitrary ways together with appropriate measuring equipment also enables side channel attacks like EM-analysis (see e.g. [AAC⁺03] for more information about side channel attacks). These attacks reveal information that could not have been obtained on protocol level.

In sum, Mallory can ruin all the goals presented in section 3.3 if no adequate security measures have been set-up. Nevertheless, her capabilities are none that are out of reach of an attacker. Only simple equipment and suitable software is required to perform the malicious actions. Thus, all RFID systems should be guarded against an attacker with Mallory's capabilities.

PHYLLIS: Phyllis is the most powerful attacker. One can expect that she has all the capabilities that have previously been presented. In addition, Phyllis has physical access to the microchips on tags. This way, she can extract data that is not revealed due to ordinary communications protocol operations.

Physical attacks on microchips have been extensively studied in the scope of smartcards because smartcards can have complex microchips that perform cryptographic operations depending on secret keys that are never revealed to the outside on a legitimate way. Thus, the development done for smartcards applies to RFID tags accordingly (see e.g. [Wei03] or [ABKL91]). Strong cryptography is not only costly to implement but offers aggressors many opportunities for attacks [Wei00].

Due to her power, all threats presented in section 3.2 are relevant regarding Phyllis, and she can ruin all the goals presented in section 3.3. However, using Phyllis capabilities requires special equipment and knowledge that is not widely available. Thus, an attacker with Phyllis's capabilities will only get into operation if the incentives are high enough, e.g. getting a shared key that compromises a larger part of an RFID system.

Simplified categorization of attackers

The categorization presented in the previous subsection is rather detailed compared to the requirements in practice. Therefore, in the following a simplified categorization which is depicted in figure 3.3 is presented.

In practice, one needs not differentiate between Tracy, Tracy-Magna, Eve, and Eve-Magna. They are similar in what they do: They all need to sniff at the air interface between readers and tags. One should not rely on the fact that sniffing is not equally difficult in both communication directions. It is just a matter of equipment to get the data in both directions so that one should see it just as an optimization of protocols not to send important data on the stronger link from readers to tags. Vicky's capabilities can be presumed for anybody in practice. Consequently, *Eve-Magna* is representative for *Vicky*, *Tracy*, *Tracy-Magna*, *Eve*, and *Eve-Magna* in the simplified model.

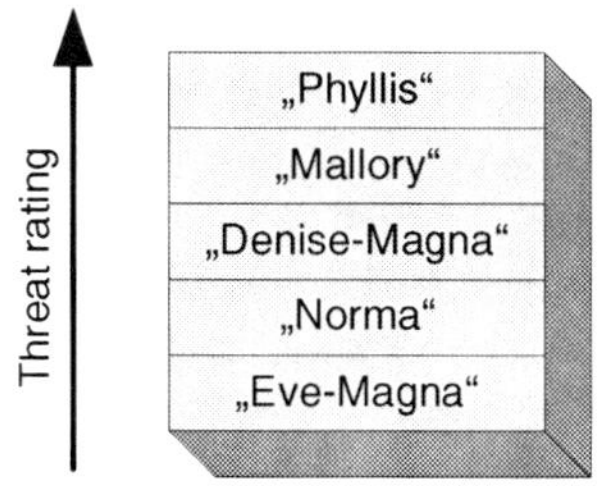

Fig. 3.3. Simple categorization of attacker capabilities

Denise and *Denise-Magna* can be combined to *Denise* since both disturb normal operation of an RFID system. The other attackers *Norma*, *Mallory*, and *Phyllis* remain as presented previously.

In figure 3.3, the attackers are aligned with an increasing threat rating from bottom to top. This means that for example an attack of *Eve-Magna* has usually less impact than one of *Mallory*. Note that the order is not absolute. Depending on the RFID system and the particular attack, the order can be different. For instance, an attack of *Eve-Magna* might be more effective than one of *Norma* in such a case. But on average, the power of the attackers increases from below to the top as shown.

3.6 Attacks on RFID Systems

Attackers can perform a variety of attacks on RFID systems. The threats from a user's point of view have already been presented in section 3.2. To adhere to the goals stated in section 3.3, one needs to keep the impact of attacks as low as possible.

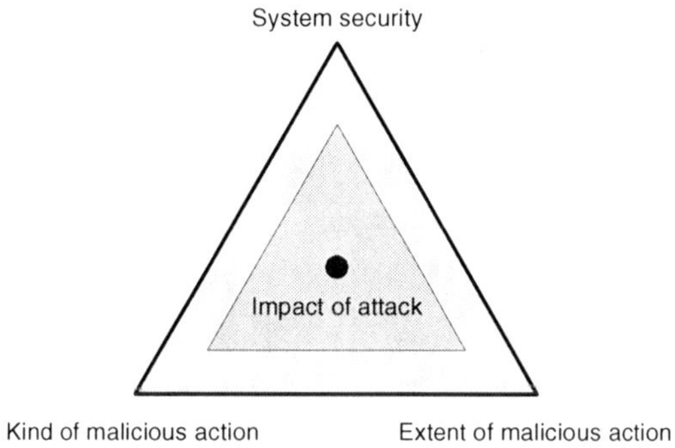

Fig. 3.4. Impact of attacks depending on three parameters

In figure 3.4, three parameters that define the impact of an attack to an RFID system are shown. The *kind of malicious action* is obviously a relevant parameter.

The capabilities of the attacker that have been presented in the previous section define the kinds of action that are possible within a certain attack. The *extent of malicious action* defines how long the attack lasts or at how many locations it takes place. For instance, it makes a difference whether eavesdropping can only take place at one location once or at several locations for a longer period of time. Finally, the *system security* defines the potential impact of an attack: If a system is well designed, i.e. it meets the challenges stated in section 3.4, the impact of an attack can be kept low even if an attacker with powerful capabilities is present and can attack for a longer time on several locations.

Categorizing attacks

In the section 3.5, the capabilities of different kinds of attackers for performing attacks have been presented. In the following, attacks shall be categorized on a higher level of abstraction.

PREREQUISITES: Depending on the attack, a number of prerequisites for performing that attack needs to be fulfilled. This includes that the attacker has the required capabilities and the appropriate equipment. In addition, he must be able to perform the attack in the required extent.

EFFORT REQUIRED: For performing an attack, an attacker needs to spend some effort. A goal of RFID system design is to increase the effort that is required for an attacker to perform a successful attack as much as possible. For instance, the effort for overcoming authentication mechanisms, e.g. by brute force, should require a high enough effort. Ideally, the required effort should be higher as the incentive of an attacker to perform the attack.

DETECTABILITY: It makes a difference whether an attack can be detected or not. If an attack is not detected, there is no possibility to catch the attacker and the system might even be in an illegal state of operation for a longer period of time. But if an attack is detected, ideally in real-time, one is able to react to the attack: At least, one can bring the system into normal operation again. If even the attacker can be caught, his actions can be sanctioned. Such pending sanctions decrease the incentive for an attacker to act against a system.

POTENTIAL IMPACT: Different attacks usually have a different impact on the given system. Thus, attacks can be categorized by their potential impact on the system. If all conceivable attacks are considered and the costs of system malfunction can be quantitatively specified, one can calculate the risk according to the risk equation introduced in subsection 2.2.2.

The criteria introduced in this subsection enable a qualitative rating of the probability of attacks if additionally the incentives for an attacker to perform these attacks are known.

3.7 Current Situation

In the previous sections 3.2 and 3.3, the threats and the goals within RFID systems have been presented. An RFID system needs to withstand a variety of attackers (section 3.5) that can perform a variety of attacks (section 3.6). Despite all the (mostly unsolved) problems, a large number of RFID systems is already in use today – usually with the intent to increase productivity.

Many of the RFID systems today are closed-loop systems that are only used within a single organization. Within such controlled environments, reaching the goals presented in section 3.3 is done implicitly: The walls of the company secure the system.

With the advent of new applications, e.g. in libraries, the technology moves to the customers. It no longer operates in a controlled environment so that security and privacy become to be at risk. Besides from some concern of privacy advocates, this is usually ignored today because the number of RFID tags and readers is still too small for abusing the technology economically, e.g. for creating customer profiles.

3.7.1 Regulation Approaches

As there are currently no technical solutions to all the security and privacy issues of the RFID technology, voluntary self-regulation is regarded by companies as a cost-effective way to cope with consumer fears.

One of the first and best known proposals for a voluntary framework for *fair information practices* is the "RFID Bill of Rights" [Gar02]. It is based on the *Code of Fair Information Practices* [HHS73] developed by the US Dept. of Health and Human Services in 1973. The five articles proposed in the RFID bill are: (1) "The right to know whether products contain RFID tags", (2) "the right to have tags removed or deactivated" upon purchase of these products, (3) "the right to use RFID-enabled services without RFID tags" (i.e. right to opt out without penalty), (4) "the right to access an RFID tag's stored data" along with the possibility to correct and amend that data, and finally (5) "the right to know when, where, and why the tags are being read".

All these articles are desirable as they would strengthen the position of customers. But even if all companies adhered to this framework, security would not be ensured and privacy would only be protected in a limited manner: People need to become active to protect their privacy. A better solution would be "privacy by default" in which no action needs to be taken explicitly to protect one's privacy. Besides that, the bill does not regulate what is done with the data after it had been accessed and with the information that is won by linking the tag data with other data about persons. Extern attackers cannot be addressed by self-regulation.

Besides the stated problems, one cannot expect that all companies adhere to such a framework because the incentive to do so is too low. The inadequacy of self-regulation approaches for the protection of privacy has already been discussed in subsection 2.3.7.

3.7.2 Assessment of EPC and Gen II Tags

GS1, an international organization dedicated to design and implement global standards in the supply chain, promotes the Electronic Product Code ("EPC"; see section 2.1.6 for a short overview) and defines requirements for tags (the current standard is called "Gen II"). This is the only widely supported approach to defining inter-organizational RFID systems that affects consumers that exists today. In this subsection, the proposed solution is discussed with regard to the goals presented in section 3.3.

EPC tags solely contain a unique identifier, the *electronic product code*. It is a superset of several numbering schemes. For the point-of-sale, manufacturer, product type, and an identifier for distinguishing individual items of a product are contained. This one will reach the consumers in the future and is thus considered here. For other identifier configurations, the considerations apply accordingly.

The vision is that all objects are equipped with an RFID tag in the future. Within current standards, the tag data is public and can be read by anybody with a conforming reader. Thus, privacy advocates fear the development.

As already shown in an example about clothing manufacturers in section 3.1, affixing tags on everyday objects can have severe privacy implications. These implications are often not obvious: Often, there are many single records of data each of which is harmless. But in combination, the data can provide detailed information about persons and their way of life.

Imagine a person carrying some items in a bag. Scanning its content with an RFID reader when passing by could reveal in the future how much money the person carries (European banknotes shall be equipped with RFID tags in the future [JP03]), what kinds of books the person reads (many libraries use RFID already today [Lin03]), what pharmaceuticals the person carries (the company Pfizer already uses RFID to combat counterfeiting of its product "Viagra"), or what personal items are in the bag.

Single pieces of information are harmless: Each item in the bag has an RFID tag affixed – no problem. Each RFID tag contains manufacturer information and a product type in an identifier – no problem. A person carries a set of items in a bag – no problem. The identifier on the tags can be read wirelessly without line of sight unnoticedly – no problem. But all these pieces of information and possibilities together lead to the privacy implications shown in the previous paragraph.

The only possibilities to prevent the privacy implications in this example are to prevent the tags being read-out wirelessly and unnoticedly (this could be done using bags with a metal mesh that acts as Faraday cage) or to remove the data about manufacturer and product type from the tags.

Kill approach

Instead of shielding the tags, one can also destroy or deactivate the tags when they are no longer required. This is known as "Kill" approach and promoted by GS1 to protect the privacy of consumers. In practice, each tag contains a 16 bit (in other literature 24 bit is stated) kill password. If a kill message with this password is sent to the tag, it can no longer be queried. As alternative approach with the same result, there are tags where the antenna can be removed manually so that the tags can no longer be queried.

The Kill approach is simple, and it is transparent for consumers how it works. But unfortunately, the approach has multiple drawbacks. One problem is that it is not applicable everywhere. For instance, libraries cannot destroy the RFID tags on checkout because they need to remain intact as long as the corresponding book belongs to the library. Identification documents and ID cards need an intact RFID tag, too. Even for supermarket items, destroying the tags at checkout is no long-term solution because useful applications are no longer possible. For instance, "smart" home appliances are envisioned: A washing machine that checks whether all the clothes in it can be washed together and that selects the appropriate program; a microwave oven that automatically selects the cooking time for an instant meal; a medicine cabinet that checks for pharmaceuticals that get too old or when it needs refilling. Another problem is that it is not obvious for consumers whether a tag is destroyed or not. Consumers thus need to trust the vendors that they perform the necessary steps for protecting consumer privacy. If consumers need to disable the tags by themselves, this is not a convenient approach: Then the consumers need to become active to protect their privacy. It would be better if privacy was protected without explicit action. In [JRS03] additional arguments agains the Kill approach can be found.

Consequently, the Kill approach appears like a work-around. It has the potential to increase the consumer acceptance regarding RFID technology, but it is no solution to the problem. Thus, other approaches are required in the long-term. In [JRS03] the Kill approach is also not regarded to be a fully satisfactory solution, "it seems imperative to explore alternative approaches".

Variants of the Kill approach like removing only the serial number parts of the identifiers and leaving the product type intact at checkout do not break the supply chain completely and prevent tracking using the unique identifier. But they are also not a satisfactory solution since for many applications uniqueness is required, tracking by constellation is not prevented, and many of the drawbacks stated above regarding the Kill approach still hold.

Track & trace

Current EPC tags do not only contain a unique identifier but also additional logic. Thus, they cannot be cloned easily. But they have by no means protection against mimicking and are thus not suited for many kinds of applications in which mimicking must not be possible. Thus, EPC tags do not include explicit measures to combat counterfeiting. It obviously takes more equipment for mimicking an RFID tag than for copying a barcode, but performing such an action is still way too simple to provide security. The EPC on a tag can be easily read out and applied to another tag or a mimicking device.

The current proposal promoted by GS1 for combating counterfeiting is *track & trace*, see [STF05]. The idea is to maintain an item history along the supply chain. This means that read-outs by actors along the supply chain are recorded centrally. Based on this, one assumes that an item is genuine if it can show up a valid, i.e. complete and reproducible, item history. Thus, track & trace can be regarded as a plausibility check.

The approach has the advantage that no special functionality is required within the tag: The unique EPC of a tag is sufficient. The first drawback is that a legitimate item and a forged item cannot be distinguished when they are read out at the same location. There is no special mechanism to prove validity. A second drawback is that the tags need to be read out regularly along the supply chain so that a complete trace of item movement can be reconstructed. But today, RFID readers are by far not ubiquitous enough to accomplish this. The third drawback is that the item history needs to be kept in a database. The data stored there reveals customer relations along the supply chain which is often not in the interest of companies. For instance, it is stored from which wholesalers a tradesman gets its goods. A fourth drawback is that the system can be fooled by infiltrating false data. For instance, valid EPCs can be obtained by reading out the tags on products within a truck. Then these tags can be mimicked at another location that appears to be a reasonable destination. If the original tags are then read out at their destination, they all appear to be forged since their location does not fit within the item history. Such attacks can be performed to unsettle companies and consumers. The fifth drawback is that track & trace is only applicable in supply chain applications. The approach cannot be used in other areas of applications, e.g. in access control systems.

Hence, track & trace does not provide an adequate solution to the counterfeiting problem. Track & trace is difficult to implement along a whole supply chain and has other negative characteristics as presented in the previous paragraph. The approach can be improved, e.g. by randomizing identifiers so that attackers cannot guess valid EPCs. Such enhancements make the system more complex but do not solve the problem because reading out valid EPCs along the supply chain cannot be prevented.

Besides track & trace, an alternative approach to combat counterfeiting using ordinary Gen II EPC tags in supply chain applications is conceivable. In [Jue05] Juels proposed to use the kill password for checking the genuiness of a tag. To prevent the tag from actually being killed, he proposed to reduce the tag powering. This procedure is not reliable and cannot be expected to work with all tags since it is not backed by the specification. A second proposal requires enhanced tags that have a restricted memory area. The access password of the tag grants access to that memory area. Product authentication can be performed by issuing the access password to the tag and checking whether a value stored in the restricted memory area equals the one stored in a database for that specific product. The problem with this approach is that it provides reliable security only once. The access password and the secret value can be eavesdropped by an attacker or be recorded by a malicious reader performing the check. After the values are known, the tag can be cloned.

Conclusion

In sum, the currently proposed GS1 approach does not reach the goals stated in section 3.3. Data security is not maintained because the EPC usually contains too much information which is given away to arbitrary readers. Counterfeiting is combated with the track & trace approach for supply chain applications. But counterfeiting cannot be effectively prevented. There is no means for controlling access to the RFID system. This enables infiltrating incorrect data. Unwanted recognition and tracking of items and thus persons is not prevented. Besides the Kill approach as workaround, there are no measures taken in this direction. Coping with denial-of-service needs to be performed by the individual applications.

3.8 Assessment of RFID Security and Privacy

Based on the discussion within the previous section, in this section a conclusion regarding the state of RFID security and privacy is drawn.

Discussion regarding the technology between different parties is difficult. The topic is sophisticated and requires knowledge in a variety of fields (see section 3.4). The conflicting interests, e.g. of consumers and other parties, makes solving the problems associated with the technology difficult.

Many scenarios that are stated by privacy advocates regarding the dangers of RFID technology currently appear far-fetched. One reason is that these scenarios often presume a ubiquitous application of the RFID technology which is not given today. Another reason is that the dangers are invisible and inconspicuous. This results in a perception of risk that does not correspond to the real risk. The perception of privacy has already been discussed in subsection 2.3.6.

Albeit the RFID not being ubiquitous today, one can expect that this will change in the years to come. Tags and readers will become pervasive. Beyond this, the systems will become networked so that a data exchange between multiple organizations can be performed ("inter-organizational RFID systems"). With an increase in applications and in relevance of the RFID technology, the number of attackers will also increase. Then the fears of privacy advocates might be justified.

In contrast to other technologies, an opt-out by individuals will not be possible in the future. Tags will be that ubiquitous that nobody can elude without becoming a hermit: Tags will be affixed to many items like passports, credit cards, money, books, clothing, and many more. This poses a severe threat to self-determination as long as the technology poses unsolved security and privacy problems.

Currently, the problems that will emerge in the near future are not addressed properly which has been shown in the previous section. The consequence is that adequate solutions need to be found and standardized quickly. The research whose results that will be presented in chapter 7 aims at bringing forward the state-of-the-art in this field.

3.9 Summary

In this chapter, security and privacy in the area of RFID technology has been discussed. By means of examples it has been shown that the technology poses security and privacy risks that need to be addressed.

RFID systems are exposed to a variety of threats like illegitimate reading of data, eavesdropping of data, copying/cloning or mimicking of tags, unwanted recognition and unwanted tracking of tags and thus objects, and malign actions causing malfunction.

From these threats five high level goals have been derived that "good" RFID systems should meet: Maintaining data security, preventing counterfeiting, prevent illegitimate access to the system, preventing unwanted recognition and tracking, and coping with denial-of-service.

Reaching all these goals is difficult since there is a variety of challenges. In this context, the conflicting objectives between different parties have been discussed as well as the multidisciplinarity of the problem. Further, requirements like scalability, inter-organizational operation or low costs have been presented.

Afterwards, security and privacy have been examined from an attacker's point of view. Attackers have been categorized and their capabilities presented. It depends on these capabilities as well as on the extent of malicious actions regarding time and place of attacks how dangerous an attacker is. Depending on the overall system security, a certain attacker may have different impact on the system.

After having studied security and privacy in RFID systems in general, the state in currently used RFID systems has been presented. The assessment of the current situation has shown that the measures taken in current systems are not adequate to the demand in the near future. Thus, in the next chapter, solutions to the open problems presented in this chapter will be discussed.

4

Securing RFID Systems

After the discussion in the previous chapter regarding the various methods for attacking RFID systems, approaches for securing the systems shall be examined now. The derived goals for secure and privacy-aware systems show the overall direction to go. Following the state-of-the art in the literature, in this chapter, the focus is laid on securing the communication between tags and readers and to ensure data security and location privacy.

In the literature, a variety of protocols has been proposed to achieve these goals. Avoine maintains a website that aims at listing all relevant works regarding security and privacy in RFID systems, see [Avo07]. Already more than 170 works are listed there currently. Many of these works propose flawed protocols that should not have been published under strict review so that others write papers about security vulnerabilities in these proposed protocols. Many others only provide incremental improvements to other works, examine only special aspects of the issues, concentrate on particular application scenarios, or evaluate other works. Thus, despite of the amount of papers listed, the community providing original and innovative concepts and protocols can be regarded rather small. Avoine and Oechslin stated in [AO05a]: "In spite of the huge interest that RFID technology has caused (and the fear of consumers), relatively few people have worked on such protocols." This statement still seems to hold true.

This chapter does not aim at repeating a presentation, comparison, or evaluation of the published approaches. Such tasks are, for example, already performed in [LSMF06] with focus on authentication and for more general approaches in [Avo05a]. Overviews can also be found in [GJP05], [Jue06], and [KEB^{+}07].

In this chapter, general considerations regarding the possibilities are taken: The main concepts that are required for reaching the security and privacy goals for systems in the considered scope are classified and presented. Existing work is filed into the classification whenever appropriate; but it is not explained in detail.

The chapter begins with a section on data management in which considerations regarding the place of data storage are taken. Afterwards, the security and privacy goals which have been derived in the previous chapter have been discussed again with the aim of finding entry points for developing solutions. In the successive section, the functionality that is required in tags for reaching the goals is presented. After some general considerations regarding the implementation of such functionality, the required functionality and the concepts for its implementation are discussed in detail. The considerations are completed by the presentation of additional building blocks that can be employed for creating secure and privacy-aware RFID systems. Evaluation criteria for RFID systems are presented before this chapter closes with a detailed discussion of a comprehensive protocol that implements the required functionality.

4.1 Data Management

In section 2.1.6, it has been presented that there are two alternatives where data is kept: Either directly on the tags or within the backend infrastructure. With the information given in that section and with the examples shown in the previous chapter, it becomes obvious that data should be kept in the backend infrastructure whenever possible.

According to the usual way of thinking, one would prefer keeping data on tags in many applications: If there are objects with tags affixed to them, it is the straightforward solution to keep all data regarding this object within the tag directly on the object so that it is available wherever the object is.

But keeping the data associated to a tag in a database, just like it is done with barcodes that link to the price of items in a database, gives much more flexibility. Data can be changed and queried without the tags being present, additional data can be augmented flexibly, and there are much more resources for data storage, data management, and data conversion in the backend infrastructure. See section 2.1.6 for a more comprehensive listing.

Also from a security perspective, not to keep data on tags is highly appreciated. As stated in the previous chapter, data is not secure in tags. *Phyllis* can extract data by physical means including side-channel attacks, *Mallory* can exploit protocol weaknesses, and effort needs to be spent to protect data against eavesdroppers like *Eve* and *Eve-Magna*. *Tracy* and *Tracy-Magna* can employ the data exchange of user data for traffic analysis purposes.

Using tags for identification purposes only and getting the required associated data from the backend infrastructure gives not only much more flexibility but also much more opportunities for security and privacy protection. In the backend infrastructure, there are not so tight resource constraints as in the tags. This enables the use of proven standard security protocols (like SSL/TLS [DA99]) and a much more powerful and fine-grained access control for ensuring data security.

Due to the advantages of storing data within the backend infrastructure, this proceeding is presumed in the remainder of this book. Nevertheless, for some particular applications like passports, storing data on a tag can be interesting (see section 2.1.6) to avoid a central storage. Extending the schemes shown here is basically possible:

After proper authorization, additional data that is stored on the tag may be revealed. The additional opportunities for attacks need to be taken into consideration.

4.2 Discussion of Security and Privacy Goals

In section 3.3, a number of goals for secure and privacy-respecting RFID systems has been presented. Within this section here, these goals are picked up again to identify first steps towards appropriate solutions.

MAINTAINING DATA SECURITY: By storing user data within the backend infrastructure as stated in the previous section, the goal of maintaining data security is almost reached. The only data that is left in the tags is the unique tag identifier and additional data that is used for protocol operation, e.g. state information or keys.

The mentioned keys should not be shared among several tags. The reason is that shared keys provide a higher incentive for an attacker for performing attacks than keys that are only relevant to single tags. If the incentive was high enough, even sophisticated physical attacks (performed by the attacker *Phyllis*, see section 3.5) would be worth the effort.

Ideally, the unique tag identifier should carry no additional information. This identifier can be implemented using a globally unique number that is randomly generated within the domain of identifiers. Then it acts as a means for unique identification without containing any structure like manufacturer or product information included. In addition, ideally, no additional state information or keys should be required within a tag. Unfortunately, this ideal configuration is in conflict with other security goals like the prevention of counterfeiting and the prevention of unwanted recognition and tracking. The configuration is also in conflict with scalability. More information about these conflicts will be given in the following sections. First, the other goals will be discussed.

COPING WITH DENIAL-OF-SERVICE: As it has already been stated in the previous chapter (see section 3.3), there is in principle no solution to the denial-of-service problem. For example, shielding using devices acting like a Faraday cage disrupts service. Instead, one needs the ability to detect denial-of-service and discover from it appropriately. Coping with the possibilities of denial-of-service needs to be performed. For our considerations here it is important that no additional means for denial-of-service attacks get introduced, e.g. bringing tags and backend out-of-sync in case stateful protocols are used.

PREVENTION OF COUNTERFEITING: Another goal that has been stated is the prevention of counterfeiting. This means that one needs to be able to effectively prevent tag cloning and the use of mimicking devices.

Cloning by eavesdropping the tag's communication or by querying the tag can effectively be prevented by maintaining an inner state that is never revealed to the outside world – neither directly in protocol messages nor by means of information that can be used to deduce the inner state. Such an inner state that is not revealed to the outside can be used to prove the genuineness of a tag. For checking whether a tag is genuine or bogus, a test that examines whether the inner state matches the expected one and thus is valid is needed. Of course this test must be designed in such a way that the inner state of the tag is not revealed to an eavesdropper.

One also wants to prevent replay attacks so that the test cannot be passed by an attacker that has eavesdropped the protocol messages of a previous test. With such a protection, the inner state in combination with the test can be used to prevent mimicking of a tag, for only the original tag is in possession of the correct inner state.

Protection against mimicking in practice (i.e. on protocol level) includes protection against cloning since a temporary mimicking can usually be done with less information than is required for cloning. The reason is that for mimicking a tag temporarily, usually only a subset of its complete functionality is required.

The methods that can be used to prevent mimicking are therefore also capable of preventing cloning on protocol level. For protecting a tag against mimicking, the genuineness of a tag needs to be checked on every interaction with the tag using certain tests. If temporary mimicking is not a problem within the given application, a test for genuineness that is performed only when required is sufficient. For instance, there could be an explicit check whether a tag and therewith a product is cloned or not when an item is sold, e.g. medicine in a pharmacy.

Note that besides a technical prevention of counterfeiting, characteristics of the underlying business process can be employed. This way, even absolutely low-cost tags without mechanisms for preventing cloning and mimicking might be sufficient for some applications.

A good example of such applications is event tickets with low-cost tags containing a unique identifier affixed: If the identifier of each ticket is logged at the entrance, copies can be detected. The holder of the first ticket is granted access in any case. The holders of other tickets with the same identifier are only granted access if they can additionally present a proof of purchase issued by a trusted shop. Otherwise, access is denied for them and they are asked to declare the source of the probably faked ticket in hope to track down the counterfeiter. The concept has the disadvantage that the first ticket holder is perhaps granted access although having a faked ticket. But this is tolerated since in return the approach is fairly simple. The concept works quite well in practice: People tend to stick to trusted shops and do not buy tickets from strangers because of the fear to buy a faked ticket with which entrance to the event is denied.

PREVENTION OF ILLEGITIMATE ACCESS: The prevention of illegitimate access to the RFID system is done using authentication: If each principal within the RFID system needs to prove that it is legitimate, an attacker cannot gain unwanted access to the RFID system. This way, infiltration of false data or other malicious activity is prevented. The required functionality in a tag corresponds to the functionality that is required for proving genuineness of tags to prevent counterfeiting.

PREVENTION OF UNWANTED RECOGNITION AND TRACKING: Each tag needs an identifier with which the interrogator can recognize the tag and thus associate data to the tag. But if this identifier can be eavesdropped or queried by an attacker, it can be abused for unwanted recognition and tracking of the tag, the object to which it is affixed and thus even the person who carries that object.

Here we have opposing requirements: An identifier is on the one hand needed for legitimate use, but on the other hand it provides a means for abuse. There are two classes of solutions to this conflict: One is that interrogators must legitimate themselves to the tag before the identifier is revealed. The other class of solutions uses tags that reveal their identity directly, but in a way that only a legitimate interrogator can understand.

The idea behind the class of solutions based on interrogator legitimation is that the interrogating party needs to legitimate itself to a tag before the tag reveals its identity. Obviously, the interrogator does not know to which tag it is talking so that the power of this approach is limited.

Revealing the identity in a way that only a legitimate interrogator can understand and use it for recognition and tracking is the other class of solutions. The idea here is that additional knowledge is required to derive the tag identity from the data provided by the tag. As each static identifier or data, whether it has an understandable meaning or not, can act as a means for recognition and tracking, it must not be static. Either it must change regularly according to a predefined algorithm or it must be changed regularly by the outside world, in practice by the backend infrastructure.

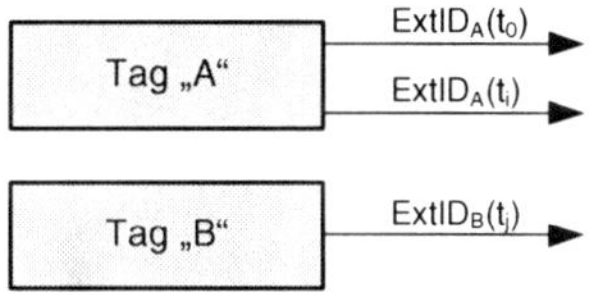

Fig. 4.1. Indistinguishability

Figure 4.1 shows the requirement of *indistinguishability* (see [OSK03]) that needs to be fulfilled so that an attacker is not able to gain usable information from the tag identifiers. If an attacker has a tag identifier of a tag A that has been obtained at a time t_0, the attacker must not be able to distinguish whether another obtained tag identifier belongs to the same tag at another time t_i or to another tag B at the same or another time t_j.

But it is not sufficient to prevent unwanted recognition and tracking on protocol level. As shown in section 2.1.4, there are multiple layers involved in the communication process. This means that tags should ideally not be distinguishable on the physical layer. This is practically impossible due to different tag implementations and variances in electrical components. Fortunately, different tag and reader alignment, different reading distances and different environments also have effects on the physical layer so that recognition just by physical layer information is only possible with special equipment and knowledge. These kinds of attacks are not considered further within this book.

Besides by the tag identifier, unwanted recognition and tracking can be performed by other means. The attackers *Tracy(-Magna)* and *Eve(-Magna)* that have been presented in section 3.5 can use traffic analysis, e.g. the amount of data that is exchanged due to a tag query, or the content of exchanged protocol messages, respectively, as distinguishing elements.

To cope with these attackers, ideally, all RFID tags should use the same protocols. Then the message exchanges have the same structure and can be implemented in such a way that for each tag the same amount of data is exchanged within corresponding protocol messages. Thus, *Tracy(-Magna)* would not have the possibility to recognize people by the tags that they carry. If all user data except the tag identifier is kept within the backend infrastructure as discussed in the previous section about data management, only the tag identifier or exchanged protocol data like state information can be exploited by *Eve(-Magna)*. Thus, protocols need to be designed in such a way that an eavesdropper does not obtain information that he could abuse.

4.3 Overview of Functionality Regarding Tags

Based on the considerations in the previous section, one notices that, seen from a high level perspective, only a limited basic functionality is required within the tags to reach the goals. Within the current section this functionality is identified and explained.

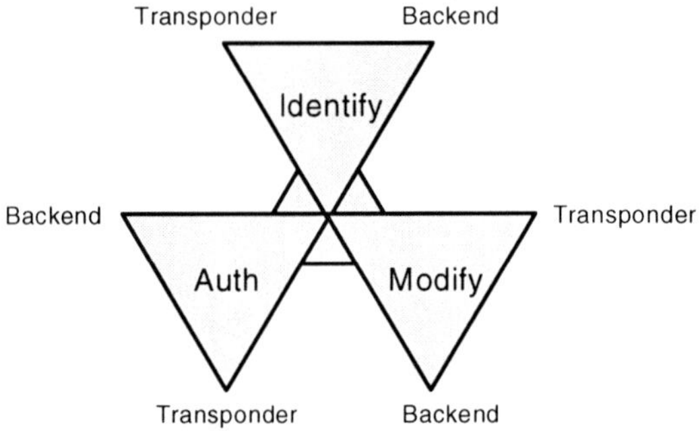

Fig. 4.2. Overview of tag functionality

In figure 4.2 the functionality that needs to be provided or at least supported by a tag is shown. This functionality can be divided into three areas, i.e. tasks: identification, authentication, and modification. Each of these three areas is divided into a part responsible for the transponder side and a part responsible for the backend infrastructure side of the communication process.

- *Identification* denotes the main functionality of a tag: It shall provide a link between an object in the real world and a corresponding object in the virtual world (e.g. a database record). This link is established by provision of a means for identifying that object automatically. The intention is to reflect the state in the real world within the virtual world to thus become able to take actions based on that state and on the detected state transitions.

- *Authentication and authorization* is the process of checking the identity of the respective principal and depending on the result behaving in a certain way or granting to perform certain actions. In the following, it will only be talked about authentication albeit sometimes authentication with a subsequent authorization is meant. For example, authentication of a tag is required to prove the tag identity in order to recognize counterfeiting. In another example, authentication of the backend is required to decide whether access to the tag identifier is granted or not (authorization).
- *Modification* is required for changing identifiers regularly. As mentioned in the previous section, doing this is a viable approach to prevent recognition and tracking of tags by illegitimate outsiders.

Identification is a required functionality for a tag to fulfill the intentions why the RFID system has been set-up. But as identification also can enable unwanted recognition and tracking, this functionality needs to be limited to legitimate principals. *Authentication* is mainly used to prevent forgery but can also be used as a prerequisite to be allowed to perform certain operations. Authentication is thus implemented for security reasons. *Modification* changes identifiers regularly to prevent unwanted recognition and tracking. Thus, modification is implemented only for privacy reasons.

4.4 Implementation Considerations

Before a more detailed discussion of the functionality that is relevant to RFID tags is performed within the next section, in the following some considerations regarding the implementation are taken. The following subsection deals with the limitations that are relevant regarding the implementation of the functionality in RFID tags. The successive subsection 4.4.2 presents the primitives that can be employed.

4.4.1 Limitations for Implementation

Implementing the stated functionality is not a simple task. RFID tags and their applications have a number of constraints. Some of the challenges have already been highlighted on an abstract level in section 3.4. Within this section here, limitations that are relevant for the technical implementation are considered.

RESOURCES: High quantities of tags are required for tagging each object within the supply chains. Thus, a low cost per tag is of special interest. This means that computational capabilities and memory on tags are tightly limited. The protocols and primitives to be used for tag-reader communication should thus be carefully designed so that as few circuitry as possible is required on the tags.

TIME CONSTRAINTS: Within many applications, it is important that the read-out of an RFID tag or even a bunch of RFID tags can take place within a short time. Thus, the time required for message exchanges and calculations is an issue that needs consideration.

NO CLOCK: RFID tags are usually simple, passive devices. This means that they do not have an own clock. RFID tags thus do not have a sense of time or time spans when they are powered off: They do not know by themselves the current date and time or when they have been read-out the last time. Any timing information needs to be provided by the outside world via the reader. This information can be wrong by error or a wrong time may be supplied by an attacker. This means that timing information cannot be used in authentication protocols to defeat certain kinds of attacks like replay of messages. Thus, other methods for coping with these kinds of attacks need to be used. But note that timing information can be employed *after* authentication of a trusted timing source.

DEPENDABILITY: Besides the stated limits, there are additional requirements. An important one is dependability. Even in case of errors by coincidence or in case of errors that have been provoked by an attacker, the system must not get into invalid states. Examples for possible errors are lost or faulty messages caused by noise on the communications channel or sudden disruption of the energy supply by the reader in case of power outage or when the tag is moved out of range. The consequence is that the RFID system needs to be robust: The implementation needs to cope with a variety of possible errors.

4.4.2 Primitives for Implementation

For the implementation of the mentioned functionality, a variety of primitives exist. It makes sense to distinguish the following five basic classes of primitives:

- based on asymmetrical encryption (public-key cryptography),
- based on symmetrical encryption,
- hash functions,
- simpler mathematical functions, and
- pools of data.

CRYPTOGRAPHIC PRIMITIVES: Cryptographic primitives, i.e. encryption schemes and one-way hash functions, have already been discussed in the required detail in section 2.4. Thus, please see that section for further information about these primitives. As already stated there, the considerations within this book focus on hash functions since they are regarded to be better suited for application in RFID tags than the other primitives.

SIMPLER MATHEMATICAL FUNCTIONS: Even basic cryptographic primitives can only be implemented with a higher cost compared to simpler primitives. Thus, some researchers propose protocols that avoid using cryptographic primitives and employ simpler mathematical functions.

One possibility is to use matrix calculations. One approach using such matrix calculations is the following: Tag and backend have a matrix M as shared secret. For authentication purposes, a challenge vector $\vec{c}$ is given by one party. The other party can then prove its authenticity by calculating $\vec{r} = \vec{c}M$ and sending the response $\vec{r}$ back. The same calculation can be performed on the other side, and the result can be checked.

The security of this approach lies in the fact, that if the matrix M is not too small (degenerated to a vector), the matrix M cannot be obtained if only a single vector $\vec{c}$ and a corresponding vector $\vec{r}$ are known to an attacker. But if enough tuples of such vectors are given, the matrix M can be determined by solving the underlying linear system of equations. For example, if the matrix is quadratic $M^{(n \times n)}$, n tuples of linearly independent vectors $\vec{c}$ and $\vec{r}$ are required to obtain the matrix. If the n vectors $\vec{c}$ are written as matrix C and the n vectors $\vec{r}$ are written as matrix R, then $M = C^{-1}R$ holds and the attacker has obtained the shared secret. If the attacker takes into consideration that the domain of the vector elements and matrix elements is limited (e.g. 8 bit) and only integer values are allowed, he has additional constraints and can probably obtain the matrix M with even less tuples.

Obviously, the security characteristics of simple mathematical primitives are not as good as cryptographic ones. Thus, they may only be used for attacker models that give the attacker limited power. Whether such assumptions are valid for real-world scenarios needs to be carefully analyzed. Otherwise, the schemes using such primitives pretend a level of security that practically does not exist. In any case, breaking of such simple primitives should only put user privacy at risk and should not impose a threat to availability, i.e. of denial-of-service, to an RFID system or its parts.

POOLS OF DATA: Instead of performing calculations, one can use storage space. A pool of data that is stored on the tag and in the backend infrastructure can be used as a shared secret.

Using pools of data for checking whether transactions to be performed are requested by a legitimate user is widely used. For instance, in many home-banking schemes, the bank and the user share a list of TANs[1]. Each TAN is a one time password that is used to acknowledge that the request is legitimate. The problem is that this scheme is susceptible to man-in-the-middle attacks.

Another approach is to use a fixed set of shared values. To acknowledge a request, the provider requests a subset of these values to be revealed. Then the user needs to respond with the requested values. The idea is that the set of shared values is not revealed completely in a single transaction and that the user cannot anticipate the challenge, i.e. the values to be transmitted, of the provider.

---

[1] TAN means "Transaktionsnummer" and denotes a one-time password

As an example of this scheme, the German "SCHUFA" [2] issues a "Schufa Card" to its customers with a matrix of values. For the login to the online services, a user password and three values within this matrix are requested.

Approaches based on a pool of data have two main parameters: The size of the data pool and the amount of data that is revealed within a single transaction. The choice of these parameters is obviously a trade-off: Ideally, data out of the pool should only be used a single time but such a large shared secret is practically not feasible. This corresponds in some aspects to the one-time pad in cryptography that is information theoretically secure but not feasible in practice. The other trade-off takes place regarding the amount of data that is revealed within a single transaction: On the one hand, this amount should be large so that it could only be guessed with negligible probability by an attacker. On the other hand, it should be small so that the data within the pool needs not to be reused often. In practice, the choice of the parameters depends on the attacker model and the level of security required.

4.5 Discussion of Basic Functionality

Each of the three areas of functionality shown in figure 4.2 are on the one hand relevant to the tag side of the communication and on the other hand relevant to the backend side of communication. Thus, the following six tasks must be considered in more detail:

- Identify tag,
- Identify backend,
- Authenticate (and authorize) tag,
- Authenticate (and authorize) backend,
- Modify tag identifier, and
- Modify backend identifier.

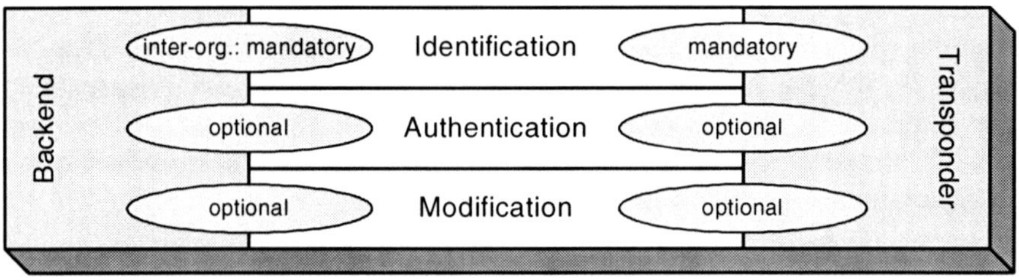

Fig. 4.3. Tag functionality requirements

In figure 4.3 these six tasks are depicted. As shown there, the identification of a tag is the only mandatory one of these tasks. The identification of the responsible

[2] SCHUFA - a company keeping data of individuals for credit assessment

backend principal is mandatory in inter-organizational RFID systems. All other tasks are optional in most protocols because they are used mainly to prevent counterfeiting and unwanted recognition and tracking which are not required for RFID system operation vitally.

The order of performing the tasks is important since different orders lead to different results. For instance, if identification of a tag was performed prior to authentication of the backend, it would not be ensured that the identification data is sent to a legitimate receiver. But if authentication was performed first, it could be ensured that only legitimate receivers obtain the data.

4.5.1 Identification

The main purpose of RFID systems is the identification of objects. This is already expressed in the acronym *RFID* itself. The simplest RFID tags just store an identifier that is sent to the reader on each query (see figure 4.4). The identifier *ID* is stored in the tag memory and given to the outside world as *ExtID* upon a reader's query. The tag identifier can be either random or consecutive or structured.

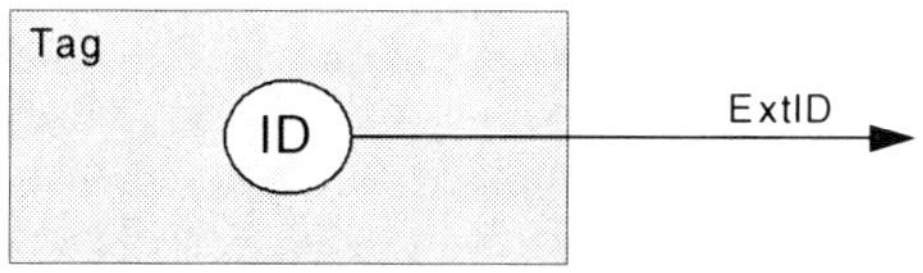

Fig. 4.4. Behavior of a simple RFID tag

A random identifier is an arbitrary value within the domain of identifiers. If the domain is large enough, e.g. 128 bit, one can expect that there will not be two tags in the world that have the same identifier. Thus, the identifier is expected to be globally unique.

If tags have consecutive identifiers, tags of a certain administrative domain have successive values. For example, an administrative domain might start numbering tags with 1, 2, 3, 4, and count on. The advantage is that the domain of identifiers needs not be as large as when using randomized identifiers. But the drawback is that tags cannot be issued easily by multiple assignment points within that administrative domain and that the identifiers are not necessarily unique across different administrative domains.

Structured identifiers are divided into certain parts. For example, an identifier could be divided into a part denoting an administrative domain and a part consisting of a random identifier. For supermarket products, a division into manufacturer, product type, and unique serial number is common. There is a variety of numbering schemes in use. The domain of the identifiers can either be divided into parts of fixed length or be used as a whole by variable length schemes. In the latter, the length of each part could for instance be defined by certain prefixes.

Using structured identifiers, hierarchies can be implemented easily. This eases administration if multiple administrative domains or multiple assignment points are involved. For example, in the case of supermarket products, each manufacturer gets a fixed prefix of the identifier space. The remaining part can be assigned by the manufacturer himself.

On the other hand, structured identifiers leak information which might be a data security or a privacy problem. For instance, if one part of a structured identifier denotes an administrative domain, an attacker learns, whether two tags belong to the same administrative domain or not just by comparing that part of the identifiers.

Identification of tag

As stated previously, the task of identification can be separated into identification of a tag and the identification of the responsible backend principal. In this paragraph, the identification of a tag is discussed.

When a tag is read, this is done to identify the tag and therewith the object to which the tag is affixed. Thus, the means to identify a tag is the basic functionality of an RFID system. In the simplest form, a tag has an identifier that is revealed to a reader within a tag query like depicted in figure 4.4.

The identifier of a tag can be globally unique. In this case, all tags world-wide can be distinguished unambiguously without using further data. In practice, the identification of a tag is only unique within the considered administrative domain. Global uniqueness is then ensured in combination with the identification of the responsible entity within the backend infrastructure that will be explained in the next paragraph.

In contrast, different objects could have tags carrying the same identifier. In this case, these objects could not be distinguished from each other just by processing these identifiers. For example, if the tag identification solely denotes the product type and does not also contain a product specific serial number, products of the same type will not be distinguishable. This case is avoided in practice by ensuring that the identifier is unique in the considered domain.

Identification of backend principal

Besides the identification of the tag, the administrative domain that is responsible for the tag needs to be identified. This administrative domain is denoted as "backend principal" or "backend entity" within this book.

For instance, in an electronic product code containing manufacturer, product type, and serial number, the manufacturer part of the identifier structure would in practice denote the responsible backend principal, whereas product type and serial number would uniquely identify the tag within the scope of the backend principal. The complete electronic product code identifies the tag uniquely, but by the manufacturer part, the responsible backend principal is denoted implicitly.

To identify the backend principal implicitly using the identification of the tag like in the presented example is common. Nevertheless, both identification tasks can also be independent of each other.

The identification of the backend principal is of special relevance in inter-organizational RFID systems. Within a single administrative domain it is not required because in that case there is nothing to differentiate. Additional information why the identification of the backend principal is required will be given in chapter 6 in which inter-organizational RFID systems are considered.

Implementation methods

There are different classes of methods with which the identification process can be implemented. The following methods have been identified:

- regular identification,
- implicit identification,
- multi-step identification, and
- encryption with shared key.

REGULAR IDENTIFICATION: The complete identifier is sent from the tag to a reader within a single logical message exchange. Note that the identifier can also be sent within the process of tag addressing in an anticollision algorithm (see section 2.1.4) like binary tree walking. In such a case, the identifier is often exchanged in several steps of communication, but this is not relevant to higher layers: After tag addressing, the complete identifier has been transmitted.

IMPLICIT IDENTIFICATION: Implicit identification is performed using information that has not been provided explicitly for the particular identification purpose. As already mentioned, the identification of the backend principal is often performed using information contained in the tag identifier.

MULTI-STEP IDENTIFICATION: In this variant, the identification is not performed within a single logical message exchange. Instead, in the first step only a part of the identification information is revealed. Only after having processed this first part and perhaps an authentication and authorization step, more identification information is revealed. This means, that the identifier is not revealed in a single step but in two or multiple ones.

Multi-step identification is an idea for solving the data security problem and the problem of unwanted recognition and tracking: Instead of revealing all information at once, only a part of it is revealed at first. More information is only revealed if the request proved to be legitimate. The problem with this idea is that the part that is revealed directly can be used for recognition and tracking by constellation. Thus, the level of security that can be reached is not very high so that it is questionable whether the advantages to be gained outweigh the disadvantages (multiple message exchanges and thus slow speed of reading).

ENCRYPTION WITH SHARED KEY: For protecting the information contained in the identifier, the identifier can be transmitted in encrypted form. As low-cost tags cannot perform the required enciphering by themselves, the identifier needs to be calculated outside the tag and then be stored on the tag directly in enciphered form.

The encryption can solve the data security problem that occurs when a structured identifier that leaks too much information is used. However, encryption of the identifier alone does not solve the problem of unwanted recognition and tracking: Enciphering of an identifier creates a new trackable identifier. This means that the ciphertext itself can be used as identifier and can thus be abused for unwanted recognition and tracking.

Besides this problem, the shared encryption key poses a problem in this approach: It needs to be properly secured because if it became public, the encryption would have no value any more. Exchanging a key that became public by a new one is practically infeasible because all tags with identifiers enciphered with such a key would need to come into the read range of a reader that can perform the change. Shared keys are thus not suited for RFID systems if long-term security is an issue [REC04].

One might ask why a shared key is used and not one key per tag. The answer is that this would not work in a scalable manner. The problem is that the principal that needs to decipher the identifier needs to know which key to use. But this depends on the identity of the tag. So there is a chicken-egg problem here: Tag specific keys require initial release of identity which itself should be protected by the encryption. Of course, the principal could try all the keys it possesses and could detect a valid one using a magic value within the identifier. But this approach is obviously not scalable.

4.5.2 Authentication

Authentication of tags and backend is used to prevent counterfeiting and also to prevent illegitimate access to the RFID system. Authentication of tags is performed to prevent cloning and mimicking and therewith counterfeiting by proving that the tag is valid. If only tags whose identity is proven may provide data to the backend, infiltration of false data is prevented, too. Authentication of the backend is important when the backend shall be able to update state information in tags. Of course, an attacker shall not be able to do so since bringing a tag in an illegal state may render it unusable. Authentication of the backend is also a prerequisite for ensuring that data is only given to a legitimate principal.

If only one of the communicating peers, i.e. tag or backend principal, is authenticated, the authentication is *unilateral*; if both peers are authenticated to each other, the authentication is *bilateral* or *mutual*. Bilateral communication can be achieved by combining two unilateral authentication processes or using an authentication protocol that combines bilateral authentication into a single process.

As shown in the previous subsection, simple RFID tags just store an identifier that is sent to the reader on each query (see figure 4.4). This completely reveals the inner state of the tag to the outside world. Thus, an attacker gets all necessary information to clone or mimic a tag just by eavesdropping a single tag answer or reading the tag himself. The situation is equivalent to an optical barcode: Just by reading the barcode an attacker has all required information to create a duplicate.

The solution is to create RFID tags that have an inner state that is never revealed to the outside world. For the authentication process, the tag must prove to the backend that it is in possession of the correct inner state without revealing it. The mentioned inner state is usually a data value. Its domain must be large enough to make attempts for guessing the value impractical.

The proof must ideally be given in such a way that it cannot be reused by an attacker later on. To prevent such replay attacks, the proof needs to be bound to certain transactions. There are different possibilities to do so. These will be explained within the considerations in the following.

The methods used for authentication can be divided into two basic classes:

- Single message authentication,
- Message exchange based authentication.

Single Message Authentication

Authentication can be performed without the creation of a communication dialog between the involved principals. This has the advantage that only unidirectional communication is required. Nevertheless, the identity of the sending peer can be proven.

Single message authentication (sometimes also called "one-way authentication") can be split into two subcategories: Authentication based on static data (i.e. a "pool of data", see section 4.4.2) and authentication based on dynamic, i.e. calculated, data.

Single message authentication based on static data (tag)

The simplest form of authentication is the provision of a password or another secret value. The password needs also to be known to the peer so that he can check whether the password is correct. As the password is a secret that is being revealed, this simple approach is obviously susceptible to replay attacks: An eavesdropper can overhear the password and reuse it later on.

A randomized identifier can also act implicitly as such a password. The "this is who I am and that's how to prove it" is then performed within a single step. The idea is here that if the domain of the identifier values is large enough, a value used within the system cannot be guessed. Then, the ability to present such a valid value is also a proof of authenticity. This approach has been proposed for making the "track & trace" approach that has been presented in subsection 3.7.2 a bit securer. The randomized identifier approach of course suffers from the same vulnerability than the described password approach: An eavesdropper can overhear the information and use it in replay attacks.

The presented scheme can be improved by using a larger amount of data. The tag might have a list of numbered authentication values. This is similar to the home-banking scheme based on TANs that has been presented in section 4.4.2, but the values are indexed. Ideally, each of this authentication values is then only used once: In the first transaction, the authentication value #1, in the second transaction, the authentication value #2, and so on. Replay is prevented by not allowing the reuse of authentication values. A major problem with this scheme is that the pool of data used for authentication values needs a lot memory for storage but nevertheless gets depleted sometime. The domain of authentication values needs to be that large that guessing an authentication value with an index greater or equal than the currently expected one is practically impossible.

For coping with the problem of depletion of a pool of static data, one can either update the data within the pool, use dynamic data, or introduce release policies for preventing depletion. As such an update of authentication data needs to be performed in such a way that an eavesdropping attacker does not gain useful information, this approach is not considered here. Using dynamic data seems to be a niftier approach and shall thus be presented in the next paragraph. Using policies is a special solution that is considered in detail in 7.1.2.

Single message authentication based on dynamic data (tag)

The idea of authentication schemes based on dynamic data is to use more powerful primitives (see section 4.4.2) than pools of data. Such primitives range from simple mathematical functions up to complex cryptographic primitives that are used for calculation of authentication data.

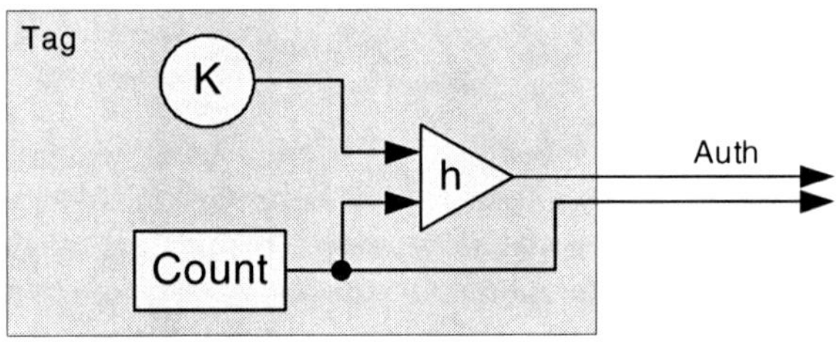

Fig. 4.5. Authentication using dynamic data with transaction counter

As already discussed, we will focus on hash functions as basic primitive. In figure 4.5 a simple approach for tag authentication is depicted. It uses a hash function as primitive and a counter.

It operates as follows: The tag contains a key *K* that is shared with the backend principal. Besides that, a counter exists that is incremented in each transaction. The current value of this transaction counter and the key are used as the preimages of a hash function. For authentication, the tag sends the resulting hash value and the current value of the transaction counter to the backend principal. The latter checks whether the counter has been increased to a new value. If so, it calculates a hash value by itself using its copy of K and the received counter value. If the calculated hash value equals the received one, the authentication was successful. Then the counter value is stored as the last seen value. Note that this operation may only be performed after a successful authentication since all transaction counter values below this value will be regarded invalid in future transactions so that a change may only be triggered by the valid tag.

A disadvantage of this approach is that the value of the transaction counter needs to be transmitted. It would not be sufficient if the backend principal operated a counter on its own and used this one as input for the hash calculation because tag and backend got out-of-sync upon a message loss or another error. So the transmission of the value of the transaction counter is required. But as the value of the transaction counter is predictable by an attacker, he could use that value for unwanted recognition and tracking.

To avoid unwanted recognition and tracking by the value of the transaction counter, one can use the current time as a means to counteract replay attacks. The principle is shown in figure 4.6.

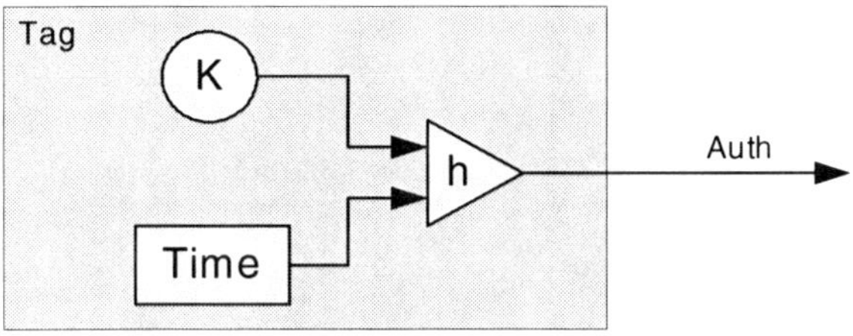

Fig. 4.6. Basic authentication using dynamic data

This method works similar to the previously presented approach with the transaction counter. But instead of using the value of the counter, both tag and backend principal use the current time as an additional input to the hash function within the calculations. Obviously, the time needs to be in sync. To cope with deviations, the accuracy of the time used within the calculation should be limited to a reasonable value, e.g. ten seconds. This way, time intervals are defined. The possibility that the backend principal is already in another interval than the tag must be accounted for. A solution is that the backend principal can also try the nearest neighboring interval after the authentication using the current interval failed.

This time based method does not have the problem regarding unwanted recognition and tracking of the approach based on a transaction counter. But in the currently presented method, there is no guarantee that the authentication value is used only once. This makes a replay attack possible as long as it takes place in the currently considered time interval, e.g. ten seconds. As a countermeasure, the backend principal could mark a time interval invalid after a successful authentication has taken place in it. But this would also prevent another legitimate authentication attempt within that interval.

As passive RFID tags do not have a clock (as discussed in subsection 4.4.1) the time based approach cannot be employed in such devices. Thus, this approach is not relevant in practice and was only presented to show that solutions that do not allow unwanted recognition and tracking exist.

After considering tag authentication, in the following two paragraphs, authentication of the backend will be considered. Backend authentication uses in principle the same concepts as tag authentication, but there are some interesting topics that shall be explained.

Single message authentication based on static data (backend)

The first approach to be discussed works similarly to the static data approach for tag authentication using a secret value, for instance a password. As it shares the same drawbacks, it is not a very interesting approach. The only way it might prove useful in practice is to use it before the tags are identified. Thus, the approach will be called *Interrogator legitimation preceding tag identification*. The term *legitimation* is used here to be able to discriminate the process from classic *authentication* which requires a preceding claim of identity.

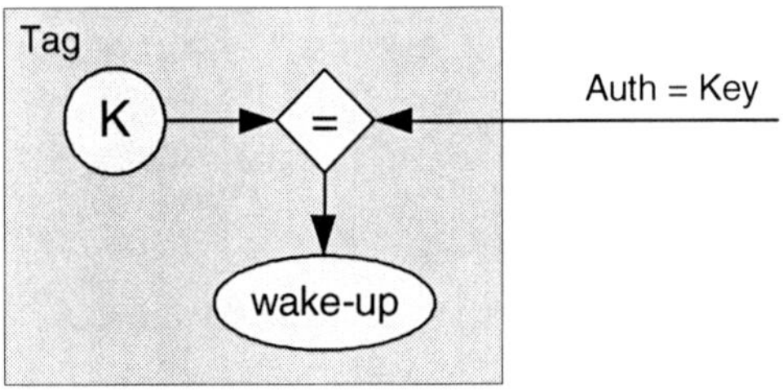

Fig. 4.7. Basic principle of interrogator legitimation preceding tag identification

Imagine a closed RFID system, i.e. one that is used for a single or limited purpose and by a tightly restricted number of interrogating parties only, i.e. usually a single organization. In this case, the basic idea behind interrogator legitimation is as follows (see also figure 4.7): After power-up of the tags, the reader broadcasts a key K, i.e. a secret value or password, respectively, to the tags. Only if that key matches a key stored in the tag, the tag takes part in the further communication; otherwise the tag goes into a sleeping state and waits for interrogation by another reader.

The described procedure is very simple and is feasible for very low-cost tags since no processing and only very few memory in a tag is required. Thus, such a sleeping state might be an alternative to the Kill approach.

Due to the sleeping state, only the relevant ones in a bunch of existing tags take actively part in the communication. This has the advantage that the process of addressing a single tag, for instance using binary tree walking, can take place with higher performance and thus the overall speed of reading is increased.

Regarding security, it is advantageous that the existence of a tag with a non-matching key is not detectable at protocol level since such a tag remains completely passive. Thus, traffic analysis, for instance for counting the total number of tags present, is not possible in this scenario. But note that passive tags do consume energy of the reader so that tag presence can still be detected by physical means below protocol level.

A problem with the approach is that the key needs to be sent in a form that can be abused by an attacker: The key can be overheard by an attacker and then be used by him for legitimating to the tags. Whether the key is sent in plain or in encrypted form makes no difference, since the encrypted key is sufficient for legitimating to the tag. In other words, the approach is susceptible to simple replay attacks.

The appealing advantages of the approach are thus contrasted by major drawbacks: The key cannot be secured effectively when sent to the tags. Besides this, the key is shared amongst a potentially large number of tags. Therefore, the solution is not very flexible and thus limited to closed systems.

Single message authentication based on dynamic data (backend)

Unfortunately, there are no powerful measures to eliminate the susceptibility to replay attacks using the approach presented in the previous paragraph. The problem is that the secret value is static. Further, initially only a unidirectional one-to-many communication is given which limits the possibilities to establish a kind of session context.

More possibilities are given if more powerful primitives are present in the tag. As already discussed in the paragraph about tag authentication with dynamic data, the widespread solution of only issuing keys whose validity is limited to a certain period of time is not viable since low-cost tags are completely passive elements that do not have a clock.

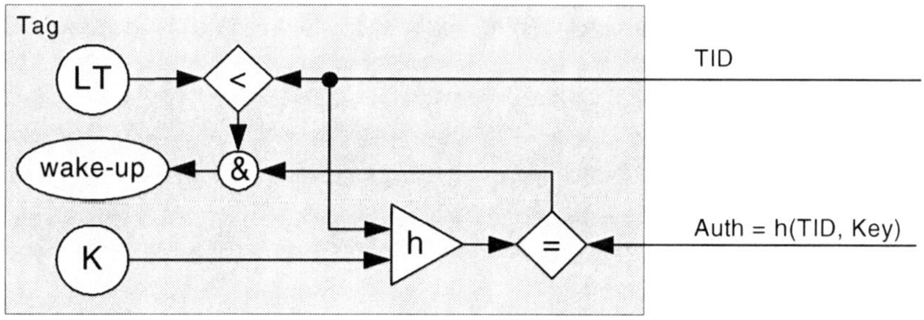

Fig. 4.8. Advanced version of interrogator legitimation preceding tag addressing

But an approach using a transaction counter is possible for legitimation of the backend principal. The process performed in a tag is illustrated in figure 4.8. The backend principal sends a transaction identifier that is incremented on every transaction. It also sends a key that is bound to that current transaction identifier. If the received key is valid, the tag stores the transaction identifier in non-volatile memory (*LT* in figure 4.8) and refuses to accept future legitimation attempts with a session identifier that is lower or equal the stored one. Using this technique, it is ensured that each tag accepts each legitimation request only once. As not all tags are read every time, the achieved level of protection is not very high: An attacker can overhear an interrogation and replay it to other tags (but to each tag only once) that did not take part in that or any future interrogation.

There also exist some additional drawbacks of that scheme: If the interrogating party used several readers, the session identifiers would need to be issued centrally so that values were used only once and in the right order. This would increases operating cost. Additionally, each tag would require additional functionality to validate the keys, for instance, a one-way hash function: In this case, the reader would calculate the hash value of the key combined with the current session identifier and send this hash value and the session identifier to the tags. The tag would check the validity of the session identifier, calculate the hash value using the received session identifier and the key stored in the tag memory and compare the result with the received hash value for key verification.

The other problem having already been mentioned is that keys are shared among a potentially large number of tags. Even if the key is not sent in a form that can be exploited in replay attacks, an attacker can get the key by physical means and this way gains access to all tags using that key. There is no way to change keys regularly if not all the tags come regularly into the read range of legitimate readers for performing a key update. Thus, the shared keys are a major drawback that limits the scheme to special applications with only low requirements.

The scheme is limited to closed systems because each party that should be able to read the tags needs the appropriate key. Giving away this key is a matter of trust since each party knowing the key can give away the key by itself.

Another problem is that in open systems usually different groups of tags exist in which each group of tags uses another key. Because of that, a reader that would like to read all the tags would need to try all available keys. With a growing number of keys to try, the approach becomes infeasible so that scalability is bad. Using the extended versions of the scheme in which hash functions need to be calculated, these calculations need to occur for each key to be tried. This takes time and resources.

The same problem of having different keys occurs when a more fine-grained access to tags is wanted: For each group of tags a key is required, and a reader needs to try all keys it knows.

In a nutshell, the presented approach has appealing features but does not scale and does not provide the security required for building open systems. Nevertheless, in certain applications – perhaps in combination with other approaches – it might prove applicable. More powerful authentication schemes can be created based on an exchange of messages. Such solutions will be considered in the following.

Message Exchange Based Authentication

In cryptography, there are multiple methods to perform authentication using an exchange of messages between two parties. Such an exchange is performed using interactive protocols. An example is zero-knowledge proofs in which a party proves that it is in possession of a secret without revealing it. Such proofs are performed in multiple rounds and are not well suited for RFID. Much more interesting are challenge-response protocols so that the following considerations will be based on them.

Challenge-response protocol basics

The idea is that one peer claims to have a certain identity. This party is called *claimant*, and this first phase is called *commitment*. Then the other peer who is the *verifier* sends a *challenge* which is usually a randomly chosen number. The claimant creates a proof for its claimed identity using the secret it possesses and the challenge. This proof is then sent to the verifier who checks whether the proof is correct. Therewith, the verifier gets to know that the claimant is in possession of the secret.

Obviously, the mentioned proof should not reveal any information about the secret. An attacker that is able to overhear the communication should not be able to get useful information that could be used as a basis for an attack. The changing challenge within each transaction counteracts replay attacks.

A good cryptographic primitive to conceal the secret which acts as preimage of the proof are hash functions. Thus, we will focus on this primitive in the following, albeit a variety of other techniques like ones based on public-key cryptography are available as well.

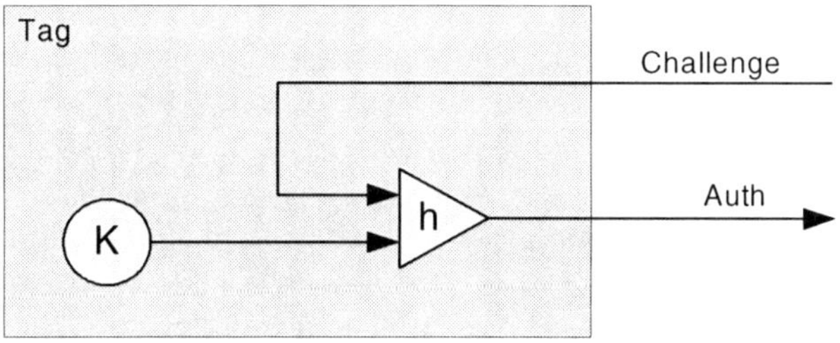

Fig. 4.9. Hash-based challenge-response authentication

In figure 4.9 a possible approach for a challenge-response protocol that is used for tag authentication is depicted. A challenge is created by the interrogator and sent to the tag. The tag combines the received challenge using a proper arithmetic operation with a static inner state *K* that is stored in the tag. The result is then sent through a hash function and afterwards sent back to the interrogator as response message.

Thereby the hash function is needed to ensure that the inner state of the tag can not be calculated by an attacker using an eavesdropped challenge and the corresponding response. The challenge needs to be different in each protocol run so that an attacker cannot use a tag answer that was eavesdropped in a previous transaction for a replay attack. A simple way for generating challenges that change in each transaction would be using a transaction identifier number that is incremented in each session. But this in not a good idea since the challenges can then be used for recognizing the backend principal. Using random numbers with a range that is large enough so that the probability for repeating challenges is negligibly low as challenge is best. Then neither the challenge nor the authentication value in the response message can be used by an attacker for unwanted recognition and tracking since both values change unpredictably.

The interrogating party needs to know the inner state *K* of the tag and can thus perform the same operations as the tag. If the calculated result matches the result that was sent by the tag, the tag passes the test and thus successfully proves its identity. So the described protocol is based on the shared secret *K*, the challenge that works as transaction identifier to counteract replay attacks, and the hash function *h* to conceal the secret *K*.

It might occur to somebody to extend the presented unilateral tag authentication scheme to a bilateral scheme in the following way: The tag authenticates itself by performing the described process. Then a third message from the interrogator to the tag can be introduced to also authenticate the interrogator. For this, the authentication value in the response message by the tag can be used as challenge message for the interrogator who then himself needs to answer with an authentication value in the new third message. The latter mentioned authentication value is calculated the same way as in the tag. But this scheme is not secure because an attacker could perform a replay attack in which he takes the identity of the interrogator. The attacker only needs to overhear a legitimate protocol run. Then he needs to give the same challenge to the tag as in that protocol run. This way, the challenge of the tag will also be the same so that the attacker can give the right response. This shows that combination of schemes can have pitfalls. In this example here, one needs to ensure that the challenge is always determined by the respective verifier.

There are many other possible approaches that can be used to validate a tag, for instance ones that are based on ciphering. Two exemplary approaches based on common building blocks that can be used for other purposes as well will be explained in short in the following.

A solution based on symmetric key cryptography can be implemented similarly to the hash-based solution. The encryption key is used as the shared secret. On interrogation, the tag encrypts the challenge using the encryption key and sends the result back. The interrogator checks the validity of the result by decrypting it and comparing it with the sent challenge or by performing the same operation as the tag and checking whether the result is equal.

A solution based on public-key cryptography is very similar to the symmetric one. The only major difference is that the tag and the interrogator do not share a secret. Instead, the tag holds a private key and the interrogator a corresponding public key. On a protocol run, the tag encrypts the received challenge using the private key and sends the result back. The interrogator deciphers the reply using the public key and checks whether the result equals the challenge that has been sent. More information regarding the use of public-key cryptography for authentication in RFID systems can be found in [Fel03].

Some pages before, the approach *Interrogator legitimation preceding tag identification* has been explained. In the following, a corresponding approach using a message exchange is presented.

Interrogator legitimation after tag addressing in closed systems

Contrary to communication in interrogator legitimation preceding tag addressing which is unidirectional one-to-many, in this approach a bidirectional communication between a reader and a single tag takes place after tag addressing.

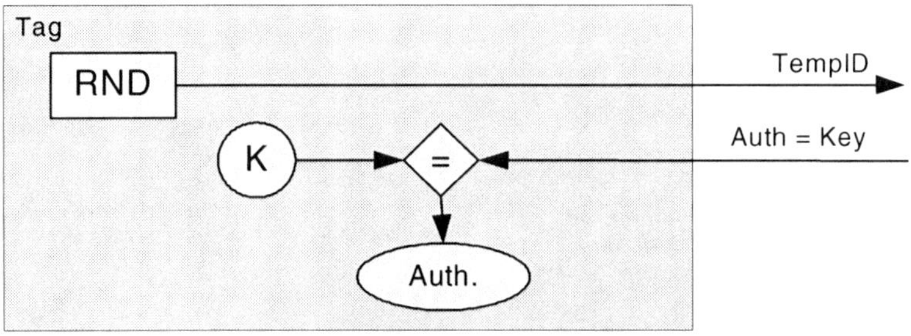

Fig. 4.10. Basic principle of interrogator legitimation after tag addressing

The basic principle is illustrated in figure 4.10. The reader establishes communication with a tag using a temporary identifier. In the next step, the interrogating party sends a key to the tag. In the case that this key matches the one stored in the tag, the tag regards the request as legitimate and reveals its real identifier and thus all information that is required for identifying and tracking the tag. If the key is invalid, the tag finishes the communication with the reader and thus does not reveal any information it can be identified with.

Temporary identifiers are a common requirement for interrogator legitimation after tag addressing. The reason is that the identifier of a tag is usually also used as physical address of the tag (similar to a MAC-address in the Ethernet) for addressing a tag by the reader. This is not possible here because the identifier may not be revealed before the communication in which the interrogator legitimates itself has taken place. Hence, another identifier is needed for use as physical address when addressing a tag within a session. This session identifier must not be static since otherwise it can be used as a means of tracking. Thus ideally, the session identifier is randomly created for each new session, i.e. for each tag power-up. Thereby, the created pseudo random numbers need to be calculated in such a way that future identifiers can not be calculated or guessed by an attacker using current and older identifiers. Similarly, it should not be possible to calculate previous identifiers by current ones. Both requirements lead to the fact that an attacker should not be able to conclude that pseudo random numbers used within different sessions belong to the same tag. This avoids the possibility of unmeant recognition.

Temporary identifiers need not be unique across multiple administrative domains or even globally. It is sufficient that with a high probability each tag within the read range of the respective reader calculates a different temporary identifier so that the tags can be distinguished. Thus, temporary identifiers can be much shorter than globally unique identifiers.

It is objectionable that in this basic solution the key is susceptible to eavesdropping and thus can be used for tag identification by an attacker. Using a one-way hash function, this drawback can be solved because we have a bidirectional one-to-one communications link between reader and tag in this scenario. The idea is to use the temporary identifier also as a challenge and to expect a legitimation message that is dependent on this challenge and the key which acts as shared secret.

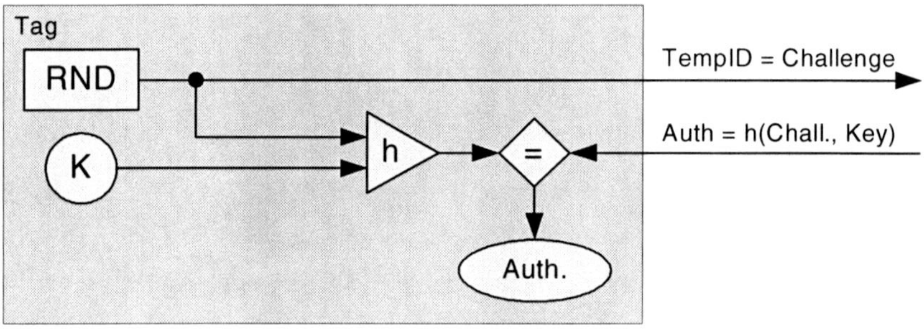

Fig. 4.11. Interrogator legitimation with protection against replay attacks

Figure 4.11 shows the described solution. To immunize against replay attacks, the temporary identifier acts as a challenge to which the interrogating party needs to answer with the result of a one-way hash function with the challenge and the key as preimages. This way, the key no longer needs to be sent over the communications medium, neither in plain text nor in encrypted form.

Regarding key sharing and key administration, interrogator legitimation after tag addressing has the same issues as already highlighted for interrogator legitimation preceding tag addressing. If the interrogating party knows several keys (because different groups of tags exist), it has to try all of them until the right one is found.

The initial message with the temporary identifier or challenge gives the interrogating party no hint of which key to use. This problem cannot be solved trivially since the initial message must not give a hint to an attacker, and there is no means to distinguish between an attacker and a legitimate reader in this stage of communication.

The advantage of interrogator legitimation after tag addressing compared to the one preceding tag addressing is the possibility to create an effective protection against replay attacks. But the price is high: The tag now needs a random number generator for the creation of the temporary identifier. Due to the one-to-one communication instead of the one-to-many communication, the speed of reading of many tags is significantly slower. As all tags take part in the addressing process, all tags in reach of the reader need to be addressed and dealt with. This also decreases speed of reading and removes the protection against traffic analysis that was accomplished by tags in sleep state in interrogator legitimation preceding tag addressing. Regarding applicability with regard to key management, the same problems apply to both schemes.

The Order of Identification and Authentication/Legitimation

The usual procedure is that at first the communicating peers claim to have a certain identity and that this claim is then proven via authentication. But this means that the identity has to be revealed at first. For staying anonymous and for the prevention of unwanted recognition and tracking, revealing the identity to arbitrary parties is not wanted.

Fig. 4.12. Possible orders for identification (1)

In figure 4.12 two possible orders for identification are depicted. On the left side, the backend reveals its identity before the tag reveals its identity; on the right side it is vice versa. In principle, there also exist cryptographic schemes that ensure that either both parties learn something or neither of the two does, but these are not relevant to RFID systems for complexity reasons and thus not discussed here.

The process on the left side of figure 4.12 appears to protect the privacy of the tag holder: The backend principal reveals its identity first, and depending on that identity, the tag can reveal its own identity or not. But the identification of the backend can be overheard by an attacker (*Eve*) and then be replayed to query the tag later on. Thus, the tag cannot be sure whether the identity information it receives is genuine or not and potentially releases its own identity to an attacker.

The process depicted on the right side reveals the identity of the tag in the first step. This can be a privacy problem because the identity information is revealed to anybody who attempts to query the tag. Analogous to the process on the left side, the backend principal cannot be sure whether the received identification information is correct or comes from a mimicking device that is operated by an attacker.

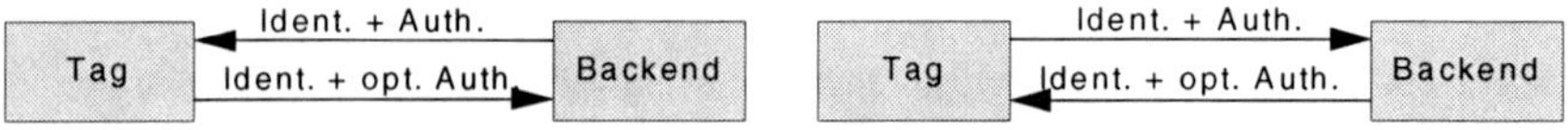

Fig. 4.13. Possible orders for identification (2)

The two presented processes show that the identification process in the respective first step should be augmented by authentication for proving that the provided identity information is valid. The two resulting possibilities are depicted in figure 4.13. Proving the identity of the respective peer in the second step is optional.

The process on the left side of figure 4.13 reveals the identity of the backend in the first step and proves the claimed identity via authentication. For the authentication to be performed without the possibility for replay attacks, the tag needs to have sent a random number as challenge before if required by the employed protocol. Depending on the identification of the backend, the tag can then decide whether to reveal its identity or not.

The process has still the disadvantage that the backend principal has to reveal its identity first. This is not wanted within many applications. If the backend principal equals the interrogator of the tag which is a straightforward approach (a more general model will be presented in chapter 6), the following example underpins this: A customer in a store, who wants to query the RFID tags of items to obtain information about the price, does not want to reveal his identity. Such queries shall be performed anonymously.

The process depicted on the right side corresponds to the previously described one. Here, the tag reveals its identity first and proves it via authentication. Depending on the tag's identity, the backend reveals its identity or not. This has the disadvantage that the tag has to reveal its identity first to whoever queries the tag. This can affect privacy because the identity information of the tag can be used for unwanted recognition and tracking.

In sum, the authentication step after the first identification gives assurance that the claimed identity is correct. But both processes shown in figure 4.13 have the problem that the initially released identity information can be abused.

A straightforward idea to cope with this problem is not to reveal the identity of a peer within a single step but within several rounds. For example, an interrogating party might first reveal that it belongs to a certain class of readers. Then the tag reveals that it belongs to a certain class of tags. If appropriate, the backend can then reveal its complete identity information, and the tag can do so as well in the last step.

But such an approach based on multiple rounds does not solve the problem completely and creates new ones. Parts of the identity information are still revealed in the first step. These are data that already can be abused. One new problem is that the correctness of the data in each step needs to be proven to counteract attackers that try to spoof the identity of somebody else. Another new problem is that multiple rounds decrease the speed of reading of the tags. Besides that, identity classes or classes of interrogator intentions or similar classes would need to be defined in a standard. It is questionable whether a generic model can be created that fits all possible application scenarios. In sum, an approach based on multiple rounds does not appear to be a viable solution.

Alternative approaches which prove the legitimation of the requesting peer before identity information is revealed have already been presented: *interrogator legitimation preceding tag addressing* and *interrogator legitimation after tag addressing*. Within the discussion, it has been shown that they are also no viable solutions for large-scale RFID systems since the scalability of the approaches is limited to small systems.

The presented legitimation approaches as well as both approaches presented in figure 4.13 suffer from the problem that the identity of the two involved peers has not been exchanged yet: Within the legitimation approaches, no information about the identity of the two peers is exchanged at all. Within the approaches shown in figure 4.13, the peer receiving the first message obtains information about the identity of the sender, but the sender does not have any information about the identity of the receiver.

In all approaches, the problem is that authentication information (or legitimation information, respectively) has to be created by one peer without knowing the identity of the other peer. Thus, the information cannot be personalized for the receiving peer. As already presented in some approaches, then only shared keys can be used or different keys would have to be tried until one matches. But shared keys are interesting targets for attackers, and revoking keys, after they have been compromised, is not possible because the tags cannot be accessed offline. Besides that, in RFID systems spanning multiple organizations it needs much trust to give shared keys to other organizations. This limits the approaches to small systems.

From the previous considerations we learn the following:

- Legitimation approaches are not an overall solution, but they can be interesting as extension.
- Authentication must be performed after identification steps before other actions on the basis of the identification information are taken.
- Identification in the first step has privacy implications.
- Performing identification in multiple rounds with tags is not a suitable approach.
- Authentication can only be implemented in a scalable manner if the identity of the peer to whom ones claimed identity shall be proven is known to the claimant.

The result is that there seems to be no appropriate overall solution. The problem is sophisticated, since the technical requirements and privacy are at odds to each other.

But a solution to this problem is possible. It consists of two parts. On the one hand, the identification of the tag needs to be performed in the first step. The privacy implications of this proceeding are solved by using identification data that has only a meaning to legitimate parties. For the prevention of recognition and tracking, the identifier is changed regularly. This functionality is introduced in the subsection 4.5.3. On the other hand, the identification of the backend principal can be omitted if it is ensured that a tag always talks to the same backend principal. This requires the extension of the current model of a tag and an interrogator/reader. The idea will be explained in detail in chapter 6. After the communicating peers are known to each other, the mutual authentication of the two communicating peers can be performed without scalability problems. Afterwards each peer can be sure that the communication partner is the one claimed, and optionally additional steps can take place.

Binding Problems

Up to now, identification and subsequent authentication have been presented. Also legitimation approaches have been considered. In practice, these tasks are not stand-alone but take place within a context.

Binding additional tasks

If additional operations shall be performed, some kind of "session context" needs to be established to ensure that the additional operations are performed with the intended peers and that they cannot be replayed in other sessions. Thus, there needs to be means to counteract spoofing and man-in-the-middle attacks.

The identification information is bound implicitly to the authentication data: The identification data is the claim that a peer has a certain identity; the authentication data is the proof. Thus, both make only sense if they belong to each other.

A simple means to bind additional operations to a session is to create a "fingerprint", e.g. using a hash function, of the requested operations and to make this fingerprint part of the authentication data. For approaches based on hash functions, the fingerprint could be an additional preimage of the hash function. The peer can now also create a fingerprint of the received requests and include it into the creation of the hash value. If the data has been changed upon transmission, the fingerprint and thus the hash value will change. Therewith, the authentication will fail. This way, the integrity of the data regarding additional operations is ensured.

Obviously, the peer cannot determine whether the provided identity information was wrong or the data regarding additional operations has changed. But this is usually not of relevance so that combining the authentication and the integrity check of data regarding additional operations makes sense in practice. Combining the tasks lessens the amount of data to be exchanged and the number of uses of cryptographic primitives. This increases the speed of reading compared to separated tasks.

Instead of using the combination of tasks, the tasks can also be performed separately. Then a session context needs to be established that binds the separate steps together. The different methods to do so are not considered here because the combination of steps is much better suited to resource scarce RFID tags.

Location binding

Besides the binding of additional operations to certain sessions, there exists another binding problem which is discussed as *relay attack* in the literature, see [KW05] and [Han05]. Usually it is assumed that the distance between tags and readers is rather low because the physical characteristics of the air interface poses limits. But one should be careful with such an assumption.

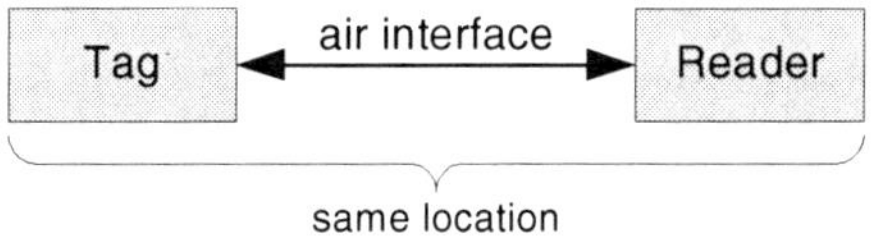

Fig. 4.14. Tag and reader at the same location

As explained in section 2.1.4, the usual read range of passive tags is limited to several meters for UHF tags or even limited further for LF and HF tags if standard-conform readers are used. Thus, if a tag is read by a certain reader, it is assumed that tag and reader are practically at the same location and not more then some meters away from each other (see figure 4.14).

A variety of applications base their operation on this assumption. For instance, a logistics company that scans the content of a packet assumes that all read RFID tags are within the packet. Another application is anti-theft protection: A shelf equipped with a reader could permanently check whether all tagged items are still there and could report any changes. Similarly, high valued items like art could be equipped with RFID tags. If a nearby reader cannot detect the items any more, one can expect that they have been moved or stolen and, in consequence, raise an alarm.

With the presented authentication approaches, it can be ensured that the tags cannot be replaced by mimicking devices. The reason is that a mimicking device cannot obtain the inner secret out of the genuine tag so that the mimicking device is not able to perform the authentication process successfully.

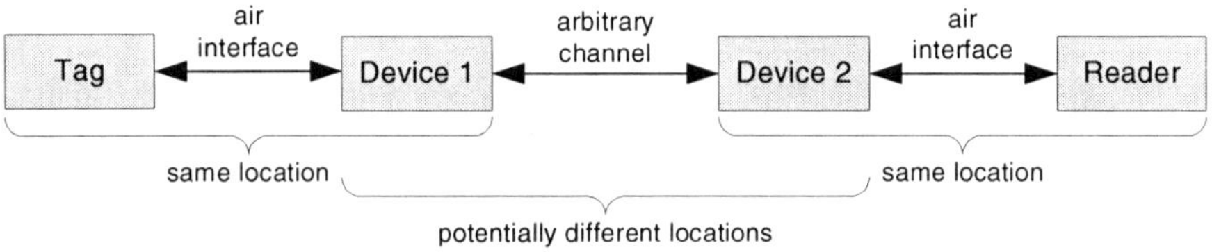

Fig. 4.15. Tag and reader at different locations

But the assumption that the location of tag and reader is always the same is wrong. Thus, an anti-theft protection or other location relevant applications solely based on RFID tag communication on protocol level do not work securely.

A potential attack is discussed on the basis of the anti-theft example in the following. Imagine the following scenario: A high value painting is equipped with an RFID tag. This tag is read by a nearby reader in the museum every ten seconds. Tag and reader perform identification and authentication tasks, so that no mimicking device can be placed instead of the tag by a thief.

Nevertheless, the thief could steal the painting without being noticed by the RFID system. To achieve this, the thief needs to place two repeater devices: One nearby the tag of the painting and one nearby the reader. See figure 4.15 for the complete installation. The repeater devices must have the ability to communicate with each other bidirectionally and act as a communication tunnel. The communication between the repeating devices could for example be realized using UMTS[3] or another mobile technology. Over this channel, the repeater devices, once activated, forward the RFID messages they receive and send the RFID messages that have been transmitted by the other repeater device.

After a query of the tag by the reader, the thief activates the signal forwarding at the repeater devices and moves the painting with the tag out of the direct read range of the reader. The next query of the reader cannot reach the tag directly any more. But the repeater device close to the reader will forward the query via the communication tunnel to the other repeater which itself forwards the query to the original tag. The answer of the tag is then propagated in the opposite direction through the tunnel to the reader. The reader thus still communicates with the original tag so that even an authentication process will be successful. But the tag is no longer physically near the reader and the thief can leave the building unnoticed by the RFID system. When the thief is away far enough, he can switch off his repeating device. The communication tunnel then fails, the reader can no longer communicate with the tag, the alarm is raised, but the thief is then already gone with his haul.

The example shows that authentication ensures communication with a certain peer. But regarding the location of the peer, no testimony can be made if only the content of the message exchange on the link and application layer (see 2.1.4) is considered.

Practical realization of the presented attack is not as simple as it is presented. The communication tunnel increases the delay in the order of tens of milliseconds. Such an additional delay would probably cause timing problems in the execution of the anticollision algorithm. But these problems are solvable: Instead of simple repeating devices, a combination of mimicking device and repeating device could be used. Operations that have a critical timing like the anticollision protocol can then be performed by the mimicking functionality locally and at the same time by a mimicking device on the other side of the communication tunnel with the peer. Authentication information and other operations that require having the correct state information can then be performed through the communication tunnel as described above.

[3] "Universal Mobile Telecommunications System", i.e. a third generation mobile communication standard

The issues for practical realization show how the problem can be addressed. One can utilize additional information besides the message content. Measuring the delay gives information whether the peer can be near or not. Another possibility is to employ other channels or lower layer information. For example, one could check optically whether the physical object to which a tag is affixed is near, or measure physical characteristics of the tag. More information on this will be given in the following subsection.

Alternative Authentication Approaches

Besides protocol based approaches, there exist other methods with which authentication can be performed or the level of security can be increased. In the following, some of these methods will be presented without going into detail.

LIMITING THE READ RANGE: Depending on the used frequency, RFID readers can have a read range of several meters (see section 2.1.4). Within this read range, not only genuine tags but also mimicking devices can be placed. One approach to increase security without introducing proper authentication is to use operation frequencies with a short read range or to limit the power of the reader for security sensitive operations. This limits the read range. Thus, tag and reader must be near each other for performing security sensitive operations.

The idea behind this approach is that physical nearness is an indication for trust. A small space around a reader is much better observable than a space with a radius of several meters so that attackers cannot hide themselves so well.

But the idea has several disadvantages. First, a short read range limits the advantages of RFID compared to other technologies like smartcards or even barcodes. Second, as explained in a previous section, location is not a priori deductible from taking part in communication: Repeater devices can be used for cheating here. A third problem is that readers or mimicking devices might misbehave. For instance, a mimicking device could bring its own source of energy, could thus receive the data from a farther distance, and send data back to the reader with a power high enough to cross the farther distance again. The gain of security by limiting the read range must thus be regarded as limited.

EMPLOYING SIDE-CHANNEL INFORMATION: Besides the content of the messages exchanged between tags and readers, there is other information that can be used for authentication or at least for increasing trust. One method is to measure physical characteristics of a tag. Each tag is unique due to fabrication tolerances. Such tolerances have measurable effects. Distinguishing these tolerances as reason for such effects from other influences like antenna alignment might be difficult, but in principle the method is feasible. Another approach would be to measure transmission characteristics like the roundtrip delay. As already explained, such characteristics can ensure that tags and readers are near each other. A high delay does not mean that the distance is large because there can be other sources of delay besides the propagation delay. But a given physical distance results in a physical bound for the propagation delay since due to the limited speed of light the propagation delay cannot be made arbitrarily small.

EMPLOYING ALTERNATIVE CHANNELS: For performing authentication, alternative channels can be used in addition to the wireless RFID communication channel. This way, the inherent weaknesses of the shared air medium can be overcome. This advantage has the price of less comfort, increased complexity or something alike. More information about alternative channels will be given in the next section, i.e. section 4.6.3.

4.5.3 Modification

In the previous subsections, techniques for performing identification and authentication have been presented. This subsection deals with the third basic task: modification.

As it has been stated previously, some kinds of identifiers are used to identify the tags and the responsible backend entity. These identifiers are thus crucial for the operation of an RFID system. But identifiers can also be used for unwanted recognition and tracking. Therefore, a solution is required that makes identification for legitimate entities possible but that does not enable identification for third parties like attackers.

Normally, identifiers are sent in clear so that the reading party directly gets essential information about the tag. A straightforward thought would be to encrypt the identifiers and only give the decryption keys to legitimate parties. Unfortunately, this approach is not feasible in practice.

Imagine an identifier *id* was encrypted with a key k. Then the ciphertext would be $E_k(id)$. Now there are two problems: (1) The ciphertext itself acts as identifier and can be used for unwanted recognition and tracking. (2) The key k must be a shared key. If it was not a shared key, the legitimate party would not know which key to use since the tag has not been identified yet so that an assignment *identity* $\rightarrow$ *key* would have been possible. But a shared key is dangerous because it would get compromised in short time as there is a high incentive for an attacker to perform an attack.

The first problem can be solved by using a ciphertext that is not static, e.g. $E_k(id, salt)$. If the salt changes, the ciphertext changes so that an abuse is no longer possible. This is a first example of identifier modification which is the topic of this subsection. By introducing the *salt*, the first problem stated in the previous paragraph is solved, but the second problem persists.

Encrypting identifiers is an obvious thought but not the right solution. But the ideas of hiding the identifier content and identifier modification lead into the right direction: (1) It should not be possible to use identifiers for identification purposes right away. It needs to be ensured that only legitimate entities can link identifiers to additional data regarding the tag. (2) Identifiers should be changed regularly so that they cannot be used for unwanted recognition and tracking. An attacker should not be able to distinguish whether two different identifiers belong to different tags or to the same tag at different times. The first stated topic needs to be addressed by proper system design: The structure of identifiers and the organization of the complete system are crucial here. The second topic can be addressed by modification.

As cryptographic operations like in the given example are very resource consuming, in [JP03] the idea has been presented to move such resource intensive tasks from the tags into extern devices. These devices shall provide a privacy enhancement by re-encrypting the tag identifier. In [GJJS04], a paper written in 2002 but published in 2004, the authors propose to use *universal re-encryption*. It is based on the El Gamal cryptosystem and makes it possible to re-encrypt a given ciphertext without knowing the key with which the encryption has been performed. But the re-encryption approach suffers from two problems: First, the re-encryption needs to be performed by trusted devices. Only these devices may alter the data on the tags. Second, the communication between these devices and tags needs to be secured. These two problems make the re-encryption idea practically useless.

In the following, the main classes of modification methods will be presented and some examples will be given. The methods used for modification can be divided into two basic classes:

- Self-contained modification,
- Message exchange based modification.

Self-Contained Modification

Methods of the first class of methods do not require a message exchange so that tags perform the identifier modification in a self-contained manner. This means that a tag can modify its identifier(s) without interacting with the outside world. In the literature, self-contained modification is sometimes called "self-refreshment" [Avo05a].

Self-contained modification takes place in connection with the identification task: An identifier is revealed to the outside world which triggers a self-contained identifier modification within the tag.

Self-contained modification based on static data

A simple method is based on static data: Tags have multiple identifiers instead of single ones. These identifiers need to be unique, act as pseudonyms and can thus identify the same tag. On each tag query, the respective tag sends another identifier out of the list. If the list is depleted, the tag starts over from the beginning of the list.

On the one hand, this method is simple and does not require computation within the tag. But on the other hand, the method is not secure since identifiers repeat. Obviously, the more identifiers are stored on the tag, the less identifiers repeat. Thus, security and memory consumption are at odds here.

A connected problem is that an attacker could query a tag multiple times until he got all identifiers. After that, he would be able to perform unwanted recognition and tracking just as if there was only a single tag identifier. Unfortunately, there is no good solution to this problem. There is no possibility to introduce a kind of synthetic rate limit beyond the technical restrictions regarding the query rate. The reason is that tags do not have a sense of time so that they cannot know how much time has passed between a current power-up and a previous one.

The only solution here is to have a static list of identifiers that gets updated regularly so that identifiers no longer repeat. For instance, such a scheme has been proposed by Juels in [Jue04]. Problems are to ensure the list update and to make such updating schemes secure against attacks. Consequently, performing such an update in a secure manner is a non-trivial task. Self-contained modification based on static data is thus not a feasible solution.

Self-Contained Modification Based on Dynamic Data

As a limited list of static data used for modification purposes is not an appropriate solution, the use of dynamic data is the next possibility. The basic idea is not to use a limited list that is memory consuming but mathematical operations to generate new identifiers.

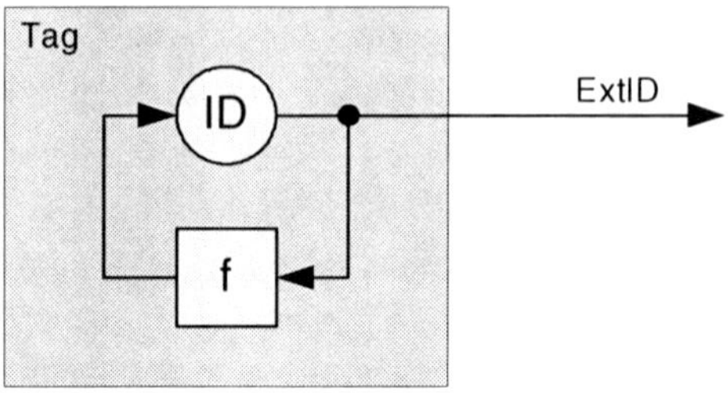

Fig. 4.16. Simple but non-secure primitive for identifier modification

Figure 4.16 shows the principle: The tag identifier *ID* is revealed to the outside world as *ExtID*. After that, a new identifier is calculated using a loop-back function: $ID^* = f(ID)$. The function needs to have particular properties for this to work: First, domain and co-domain need to be the same. Second, the domain needs to be at least as large as the domain of the identifiers *ExtID* to be revealed. If the domain of the function is larger, the *ExtID* needs to be trimmed. Third, the output of the function should be uniformly distributed so that all possible outputs appear with the same probability. Thus, a hash function would suit the requirements, but the one-way property is not required here if forward secrecy (see section 4.7 for information on that) is not demanded.

The presented primitive in figure 4.16 has two major flaws. First, if the attacker knows which function f is used, he can calculate the identifier changes by himself as well so that the scheme is rendered useless. It may not be assumed that f can be kept secret. Second, there is a synchronization problem between tag and legitimate reading entity. Both need to perform the same calculations synchronously so that the current identifier sent by the tag matches the one currently expected by the legitimate entity. But it cannot be expected that both peers stay in sync because the tag could also be queried by illegitimate readers. Adding a transaction count so that the legitimate entity knows how often the function f has been applied is not a solution because this value could be used for unwanted recognition and tracking. If an attacker performs spoofing by pretending to be the tag and then provides very high transaction count values, this could also lead to a denial-of-service because a high transaction count value would trigger many iterations for calculating f.

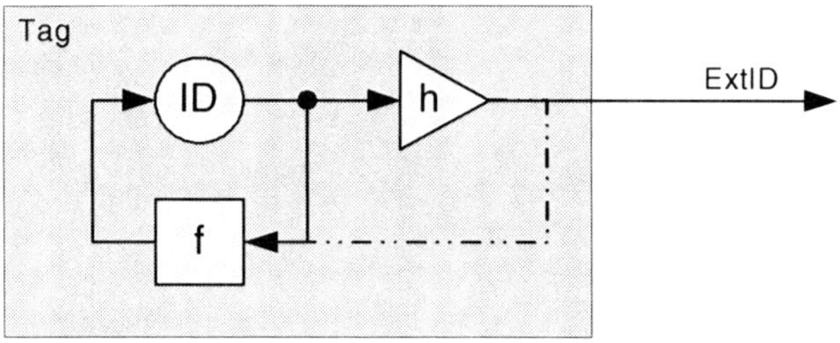

Fig. 4.17. Hash-based primitive for identifier modification

Figure 4.17 shows a hash-based solution to the first problem mentioned. It is similar to the previously shown primitive, but the internal identifier is not revealed directly but in hashed form instead.

The result is that the internal tag state is not revealed to the outside world. Therewith, it is not known to the outside world on which current internal identifier ID the function f operates. Of course, a legitimate entity needs to know the internal identifier ID so that it can perform the same operations in sync. The synchronization problem between tag and legitimate entity persists.

The function f can be replaced by a hash function. The resulting scheme is the one proposed by Ohkubo et al. in [OSK03]. It adds forward secrecy because even if the inner state of the tag becomes known to an attacker sometime, he cannot obtain previous inner states without being able to invert the hash operations. For the scheme to be secure, the presented hash function h and the replacement function for f need to be different. Instead of implementing two distinct hash functions, one can use one function with different initialization vectors or add a second preimage that is different in each case.

The dotted feedback line in figure 4.17 depicts a setup that is flawed. Here we would gain $ID^* = f(h(ID))$; and as $h(ID)$ is revealed to the outside world, an attacker can calculate the internal identifier if f is known to him.

The synchronization problem is so severe that it renders the previously presented solutions practically useless. In both methods shown, message loss or tag interrogation by an illegitimate entity can bring tag and legitimate entity out-of-sync. A precalculation of several iterations by the legitimate entity does not solve the problem because the number of precalculated identifiers is limited. An attacker just needs to query the tag often enough so that the precalculation is no longer sufficient. Then, synchronization is lost completely and the tag is rendered useless. Limiting the read rate is practically infeasible since tags have no sense of time and legitimate applications might be negatively affected. As already explained, introducing a transaction counter is no solution either.

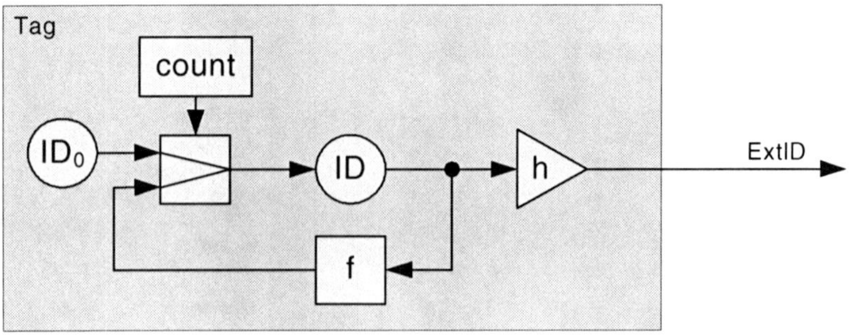

Fig. 4.18. Hash-based modification with repeating identifiers

A possibility would be to limit the number of identifiers that need to be precalculated. Then the list of precalculated identifiers needs to contain all identifiers that can potentially occur. This can be done using the approach shown in figure 4.18. The idea is that, starting from an identifier ID_0, identifiers are calculated just as shown in the previous figure 4.17. After a number $n - 1$ of calculation rounds, the ID is set back to ID_0 so that the cycle starts anew. This way, n identifiers repeat cyclically.

Therewith, the legitimate entity can perform a precalculation of the n possible identifiers so that tag and legitimate entity can no longer come out-of-sync. But now there is the same problem as in *self-contained modification based on static data*: As identifiers repeat, the security of the solution is much lower. The features are similar to the *self-contained modification based on static data*; there occurred only a shift from memory requirements towards computational requirements.

Thus, a regular update of ID_0 would be required before identifiers start to repeat. Such an update would no longer be a self-contained modification. Methods for such an update can thus be found in the next subsection about *message exchange based modification*. If such an update is implemented, it might be useful to split ID_0 into a static part and a dynamic part: The static part acts as shared secret between tag and legitimate entity; the dynamic part is updated regularly via an appropriate message exchange. This lowers the security requirements on the dynamic part: Even if that part is revealed to an attacker, the attacker cannot precalculate the identifiers because he still lacks the static part.

Figure 4.19 shows a completely different approach. The idea is here to combine identification and authentication tasks with identifier modification in a simple manner: The challenge for tag authentication is used together with the internal identifier as preimages of a hash function whose hash value is the tag identifier that is revealed to the outside. A changing challenge thus leads to a changing tag identifier *ExtID*.

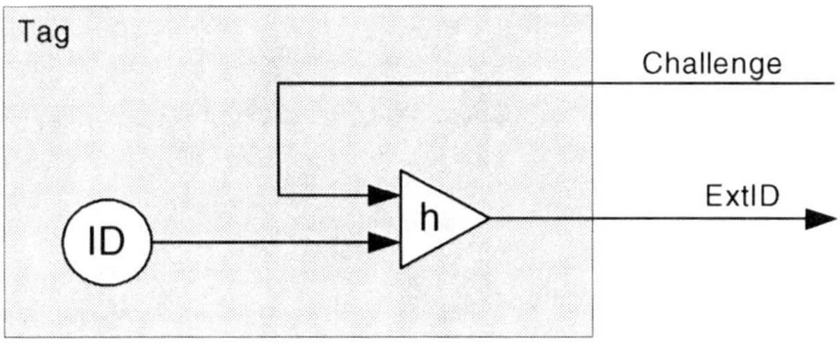

Fig. 4.19. Hash-based solution against tag mimicking

The approach is also a simple solution against tag mimicking: The internal identifier is not revealed to the outside. If the domain of the identifier is large enough and the identifier cannot be guessed, an attacker cannot mimic or clone the tag.

The shown approach has some severe drawbacks. First, it is not scalable since it has a similar problem as the randomized hash-lock approach [WSRE03]: The legitimate entity needs to perform a hash operation for all registered tags using the current challenge to generate the tag identifiers that can be expected using the current challenge. Only this way it is possible for a legitimate entity to identify the tag. A second problem is that providing the same challenge always leads to the same tag identifier *ExtID*. Thus, the scheme provides protection against an attacker like *Eve*, but not against a stronger attacker that is able to query the tag by himself and thereby can choose the challenge. Such a stronger attacker can recognize the tag.

Message Exchange Based Modification

As shown in the previous subsection, self-contained modification runs into scalability problems if security and privacy demands are not cut back. In this subsection, the second class of modification methods is presented: Modification based on an exchange of messages.

In such scenarios it is essential that the integrity and the authenticity of messages are guaranteed. The consequence is that message exchange based modification does only make sense in combination with authentication. Message exchange based modification can therefore not be presented independently from authentication.

As already shown, authentication can be performed using a variety of concepts and methods. Combining these with different possibilities of message exchange based modification results in an unmanageable number of approaches so that it is infeasible to produce a comprehensive overview.

However, the approaches for message exchange based modification can be divided into two subclasses. In the following, these two subclasses of methods are presented with a single example of each. These examples point out that designing a secure message exchange is a sophisticated task. A complete example of the combination of modification, authentication, and identification that addresses the identified problems will then be presented in section 4.8.

Send identifier update

The simplest form of message exchanged based modification is to send an identifier update in a message from the legitimate entity that is in charge of the tag to the tag. Let us assume that the origin and the integrity of this message are proven to the tag using appropriate methods.

Already this simple example highlights some pitfalls. A problem occurs when the identifier update message gets lost – whether by chance or by malicious activity. Then the entity in charge of the tag and the tag itself can get out-of-sync regarding which tag identifier is current. The usual solution would be to demand an acknowledgement message from the tag back to the entity. But this message can get lost, too, so that a retransmission scheme would need to be implemented. Similar problems are present regarding reliability of delivery in networking protocols like TCP[4]. The vulnerabilities that have been found in TCP implementation in the past decades show that designing a reliable and secure protocol is a complicated task. Here in the RFID scenario, there are additional problems: The communication link between tags and readers is comparatively slow but on the other hand, tags can only be expected to be in the range of a reader for a short time. As the energy for operation of passive tags comes from the readers, a tag does not have much power left to perform error recovery and getting into a save state after power interruption. Thus, reliability needs to be ensured albeit the environment being very unpredictable and unreliable. A possible solution to the reliability problem will be presented later in section 4.8.

Besides the reliability problem, computational resources and memory storage in tags are scarce, but the wireless communication between tags and readers takes place over an insecure channel. Securing this channel with the few available resources is a challenge.

In the example, the identifier update has been sent in clear. This means that an eavesdropping attacker like *Eve* is able to overhear the new tag identifier. Thus, the approach does not give protection against all kinds of attacks. Nevertheless, there is a positive effect in some widespread scenarios: Imagine an illegitimate stationary reader. When querying the tag, this reader gets the tag identifier. Normally, the tag identifier is the same when the tag returns to this reader at a later time so that the tag can be recognized. If the update method is employed, that reader is not able to recognize the tag if a legitimate reader has changed the tag identifier interim. But note that the reliability problem is still unsolved so that the approach is not practical without refinement.

Figure 4.20 depicts an enhanced approach that relies on symmetric cryptography. It solves the problem that the new identifier is revealed upon transmission by enciphering: The new identifier and a "magic number" *MN* are enciphered using a shared key. Now the tag can decipher the message data using its copy of the key. If the magic number is the expected one, the origin of the message is proven to the tag.

[4] Transmission Control Protocol, see [Ste94]

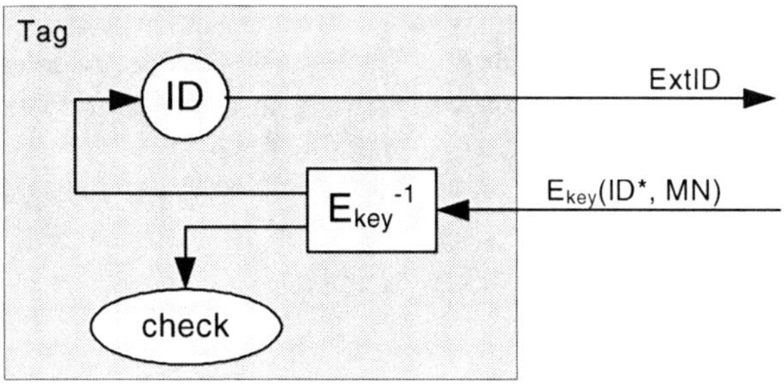

Fig. 4.20. Flawed example of identifier modification using symmetric cryptography

This approach uses a complex cryptographic primitive but is nevertheless not secure. First, the problem of message loss that can bring tag and legitimate entity out-of-sync is not solved. Second, the tag identity is not proven so that tag mimicking becomes possible. Third, replay attacks are possible: An attacker can overhear the update message and replay it to the tag at a later time. The tag in the example is not able to detect whether a certain message has already been received and processed or not. This way, tag and legitimate entity can be brought out-of-sync as well.

Send updating instructions

In many approaches shown, the intern identifier is not revealed to the outside world. Instead, the hash value of that identifier is used as external identifier $ExtID = h(ID)$. Instead of sending a message with a new internal identifier ID, one can send instructions how to update the existing one. This has the advantage that an attacker does not gain usable information by overhearing the update message. The idea has been published in [HM04b].

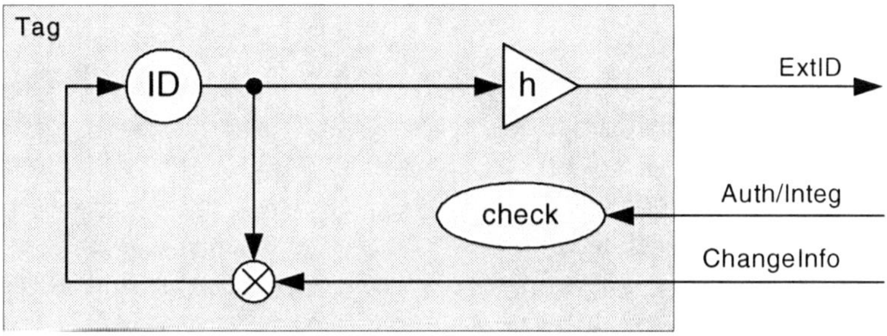

Fig. 4.21. Concept of sending updating instructions

Figure 4.21 shows an implementation of this idea. An external identifier $ExtID$ is calculated using a hash function $ExtID = h(ID)$ with the internal identifier ID as preimage. Using a non-zero $ChangeInfo$, a new internal identifier can be calculated using a xor-operator by $ID^* = ID \oplus ChangeInfo$ in which $\oplus$ denotes an xor-operator and all variables have the same domain. To make this approach secure, the authenticity of the update message and the integrity of the *ChangeInfo* need to be proven to the tag using appropriate techniques. In addition, the possibility of replay attacks needs to be counteracted and the reliability problem needs to be solved. A possible solution that addresses all these requirements and that uses the concept of sending identifier updates will be presented later in section 4.8.

4.6 Additional Building Blocks

In the previous sections, the basic tag functionality that needs to be implemented for secure and privacy respecting RFID systems has been presented. Used building blocks have been static and dynamic data as well as cryptographic primitives. The involved entities communicated via messages of data over the wireless air channel. This section will present and discuss some additional building blocks that can be employed.

4.6.1 Distinguishing Different Tag States

In the previously presented examples for the implementation of basic functionality, tags showed a certain behavior according to the specified protocol. Of course, when executing such a protocol, different states occur like "authentication successful" or "authentication not successful". However, the executed overall protocol remains the same all the time.

In [IKY02], the concept of distinguishing major tag states is introduced. Concretely, the authors propose to distinguish a public state and a private state in which a tag behaves differently. Their idea is to shadow the usual tag identifier by a user identifier. Switching between the major tag states shall be performed by using a trusted communication channel.

The basic concept of distinguishing different major tag states in which tags behave differently and in which tags might use different protocols is very interesting because it extends the possibilities for the design of comprehensive solutions regarding security and privacy protection.

Besides distinguishing a public and a private state, distinguishing a sleep and a wake state has been proposed in different varieties. A variant by the author is one for supermarket application: In the wake state, tags behave like ordinary RFID tags. If each tag has a key using which the respective tag can be put into a sleeping state, this operation can be performed at the checkout. While in this state, tags do not take part in reader queries and behave as if they were not present.

The point-of-sale terminal can print a key on the sales slip with which the tag can be reactivated. If a tag receives its reactivation code, it leaves the sleep mode and again acts like an ordinary tag. This approach is superior to the Kill approach when processing customer complaints or in cases where the customer wants to have working tags for non-supply-chain applications. The approach is simple but consumer privacy is protected.

4.6.2 Evaluating Lower Layer Information

In [FR03] the idea is stated to use lower layer information to enhance security and privacy. The authors claim that by measuring the signal strength and the noise the distance between tag and reader can be estimated. This measured distance could be used for defining a level of trust: Readers that are near to a tag a more trusted than readers farther away.

This is an interesting idea, but it needs discussion. One question is whether the assumption that a low distance leads to high trust holds at any time. Probably this is not the case, but the distance can be regarded as an indication for trust. Another question is the technical feasibility. Many aspects like tag orientation and imperfections in the reader energy field influence the signal strength and the noise, too. The distance is just one aspect. Thus, one cannot be sure that one is able to conclude the distance from signal strength and noise. This is especially the case if the energy that readers emit is not always the same. Perhaps an attacker can pretend to be nearer to a tag by increasing the energy that his reading device emits. Thus, employing lower layer information leads to many questions, but it is an interesting concept.

4.6.3 Alternative Communication Channels

Up to now, only the wireless communication over the air interface has been discussed. This neglected all the other channels that are in principle available for communication purposes. In this subsection, these alternative communication channels will be presented.

Optical channel

The most obvious alternative channel for RFID communication is the optical channel. It is already known from barcodes where the optical channel is the primary channel.

WRITTEN TEXT: Besides putting RFID labels on the packaging of an object, one can also write plain text on that packaging. This text can contain information that is not readable using the wireless RFID communication. This way, passwords or other numbers or codes can be attached to an object and can be used for identification or authentication purposes.

As written text is not machine readable in a simple manner (OCR requires clear text and proper alignment or position identification patterns but can in this case be used), usually human interaction is required. This way, the text can be used as an additional means for authentication for special operations that shall require human attention and action. For example, in [WSRE03], a master key that is printed on the interior of a package is proposed as a key recovery mechanism.

OPTICAL BARCODES: Optical barcodes are the most widespread example of auto-id systems. The optical channel is used for scanning the barcodes and therewith reading the data on them. This data can contain identifying data that is redundant to data that is present on the RFID tag. But it can also contain passwords or other data that are not contained on the RFID tag. By demanding the use of the optical channel to read out this data, one can ensure that an additional interaction with the object needs to take place besides a potential unnoticed read-out of an RFID tag. For example, if an authentication code that is contained in a barcode is required for performing an operation, it is ensured that the operation cannot be performed while the object is in a bag (of course, one needs to presume that the content of the barcode has not been revealed to an attacker before the object has been put into the bag).

Barcodes are intended to be easily machine readable and can thus be used if the increased security of the use of an additional channel shall be leveraged automatically. Using RFID and barcodes in parallel is also an interesting option if barcode applications are to be migrated to RFID. In this case, the barcodes can be used like before, optionally with additional data, until all readers of all involved parties are migrated to RFID readers. It is also possible to use RFID and barcodes in parallel for a long time. For instance, barcodes are applied to all items, and RFID is in addition applied to all items of high value. Then the genuineness of the high value items can be proven using RFID authentication, but old applications or items for which this additional protection is not economic use barcodes.

OPTICAL INPUT: Optical barcodes enable the storage of data on the respective object and the transfer of data in the direction from the object to the backend infrastructure. For certain applications, it would be helpful if data could also be transferred into the other direction, e.g. for changing memory content in the RFID tag in a more secure manner. One possibility to do so is to equip RFID tags with photodiodes. This idea seems to be unpublished up to now.

If an RFID tag is equipped with a photo diode, a device communicating with that RFID tag can use flashes of light to transfer data to the RFID tag using the optical channel. Acknowledgements indicating correct reception can be sent back using RFID communication. Using flashes of light as a secondary communication channel ensures that the communication from backend infrastructure to RFID tag takes place using a line of sight so that a manual alignment of object and light emitter is needed and the communication cannot be performed for hidden items, e.g. ones that are stored in a bag.

A photo diode can be used for communication purposes as presented in the previous paragraph, but it can also be used for obtaining state information from the outside world. For instance, light could be used as a prerequisite for the activation of an RFID tag: As long as the tag is in the dark, it does not respond to queries; in other cases, the tag behaves like an ordinary tag. Such a proceeding could be used for improving security of passports: As long as they are in a pocket or bag they cannot be queried. This would free people from having to put the passport into metal shielded bags for preventing unwanted queries.

Physical contact channel

A physical contact channel is a communication channel that has already been used for years in another context: smartcards. The advantage of a physical contact channel is that wire tapping is much more difficult than performing an attack using wireless communication with a shared medium.

The idea of using a physical contact channel for security relevant operations has already been stated in 2002 for the RFID context: "Requiring physical contact for critical functionality helps defend against wireless sabotage or denial of service attacks." [SWE02]. Before that, in 1999, Stajano and Anderson had argued for the use of a physical contact channel for pairing devices in ad-hoc wireless networks [SA99]. In that paper, they compared the pairing or initial configuration of a device with a duckling that recognizes the first animal it sees as its mother. For the pairing, they propose to use a physical contact channel as a simple and effective means for performing that action: "No cryptography is involved, since the secret is transmitted in plaintext, and there is no ambiguity about which two entities are involved in the binding." Such an "imprinting" process can in a similar manner be performed in the RFID context, e.g. for initial programming of tag identifiers or keys.

Alternative channel summary

The presented examples for alternative channels for communication show that besides the standard RFID communication a variety of alternative channels exist. These alternative channels can be used for a variety of purposes. In particular, a physical contact channel can be used for initial configuration or for reconfiguration. Alternative channels can also be used for transferring security relevant data like authentication data or encryption keys.

Besides the given examples for alternative channels, other channels are conceivable. For instance, certain environmental conditions can be used to bring an RFID tag in a certain state, for example for reprogramming: e.g. a certain level of temperature or pressure around the RFID tag or the presence of a magnetic field. It is also conceivable that only certain forms of optical excitation like by ultraviolet light force an RFID tag to enter a certain state. This way, security sensitive operations can be enabled explicitly in a form that cannot be imitated by an attacker who only has access to the wireless air channel.

Alternative channels can thus provide additional security for special operations. But the drawbacks of using such alternative channels need to be considered, too. There can be additional costs that can be near zero (e.g. printed text or barcodes) but also reach a considerable amount (e.g. additional sensors). Using additional channels makes the use of the object less comfortable because the advantages of RFID communication that takes place wirelessly and without a line-of-sight are reduced. Thus, one needs to weigh up the pros and cons of using an alternative channel.

4.7 Evaluation Criteria

Lots of different protocols for securing RFID communication can be built with the presented building blocks. These protocols are usually embedded into a large RFID system architecture. When creating such RFID system architectures, one needs to have criteria to assess the quality of different architectures and to compare different architectures to each other.

In literature, often only security aspects and resource consumption are taken into consideration. Other criteria are only mentioned when appropriate. The following presentation of criteria will highlight that there is a variety of criteria for creating a distinction between different architectures.

SECURITY: The criteria regarding security have already been explained in detail in the previous chapter. The threats have been presented and the goals that a good approach should reach have been inferred. These are within this book: Maintaining data security, preventing counterfeiting, preventing illegitimate access, preventing unwanted recognition and tracking, and coping with denial-of-service. Details have already been presented and are thus not repeated here.

Besides the high-level goals, there are conceptual and implementation specific differences between different RFID system architectures and the used protocols within these architectures. For instance, from a security point of view it makes a difference whether there are static secrets like keys in a tag that never change, or dynamic secrets that change regularly. In the first case, an attacker gains more input data to perform attacks than in the latter case.

RESOURCES: Resource consumption is a relevant criterion because it is directly associated with cost. If more resources are required, the system becomes more expensive and the economic incentive to use it decreases. The most costly resources are the ones in RFID tags because the tags are deployed in huge numbers. But resources in other devices like readers or hosts within the backend infrastructure are also of interest.

It makes sense to distinguish several kinds of resources. An important one is memory use. One can distinguish volatile and non-volatile memory here. The second important class of resources is computational resources. The resources consumed by cryptographic primitives account for the highest amount of resource use in RFID tags. The resource consumption depends on the type of primitive and the number of invocations of the primitive that is required for a single operation. A third class of resources is network resources. Network resources are relevant within the backend infrastructure. The communication of RFID tags can also be regarded as network resources.

SCALABILITY: The next crucial requirement that RFID systems intended for global use need to fulfill is scalability. For instance, the authors of the study [GIB07] predicted that until 2022 six million RFID readers would be operating at 450000 locations with 86 billion tags purchased annually and claimed these numbers to be conservative. These impressive numbers clearly indicate that large-scale RFID systems need to be well scalable.

As scalability is a crucial evaluation criterion for RFID systems, system design needs to adhere to it. RFID systems need to have an architecture that supports the partition of different system parts so that responsibility and therewith load can easily be distributed on many computers in potentially different locations. The complexity of operations should ideally grow not worse than linearly with the number of tags in the system. A negative example is the "Randomized Hash-lock" approach [WSRE03] that, if the number of tags in the system is n, requires on average $\frac{n}{2}$ hash operations per tag read. Multiplied by the number of tags, this yields a complexity of $O(n^2)$.

SUSTAINABILITY: A criterion that is seldom mentioned in literature regarding RFID system security is sustainability. Nevertheless, it is very important. The security of RFID systems is usually based on cryptographic primitives. But primitives that are regarded secure today cannot be regarded to be secure in ten or twenty years.

But the expected useful life of RFID tags is very long in some cases. For instance, tags that are affixed to books in libraries might be present and in use for decades if they still satisfy the needs. Thus, one needs to deal with long-term security of RFID tags and requires transition techniques to new cryptographic primitives in RFID systems. One can distinguish two requirements:

- possibility of multiple cryptographic primitives for the same tasks,
- keeping impact of breaking cryptographic primitives low.

A sustainable RFID system needs to be able to employ multiple cryptographic primitives for the same tasks at the same time. Imagine all current tags use a cryptographic primitive A. In the course of time, one might want to replace this cryptographic primitive by a newer, more secure cryptographic primitive B. There is obviously no way to introduce a flag day on which all tags need to use primitive B and on which A is no longer supported. The reason is that replacing all tags would be too costly. Further, there are too many tags in the wild to make a short-term replacement feasible. The conclusion is that primitive A and primitive B need to be used in parallel for a potentially long time. Old tags use an old primitive, newer tags use a newer primitive, and the RFID infrastructure needs to be able to cope with them all.

At first sight, it seems to be simple to support tags using different cryptographic primitives. Within communication protocols in networks it is common to negotiate protocol versions and peer capabilities in the connection setup phase. If an RFID tag did the same and the number of primitives in use at the same time was high enough, in many situations a tracking by constellation would be possible. One can argue that the risk is not very high, but for privacy reasons a tag should leak as few information as possible.

A second aspect is that the impact of a broken cryptographic primitive should be as low as possible. If a primitive is broken, this might render the tags in some approaches unusable, e.g. equivalent to a permanent denial-of-service. In other approaches, the tags and the overall system might be able to operate as usual and "only" the privacy protection is no longer given. Obviously, a scenario like the second one is preferable.

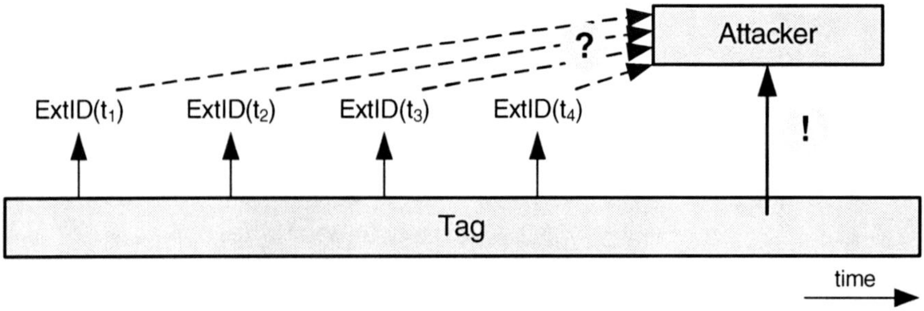

Fig. 4.22. Forward secrecy in an identifier modification example

Independently from the future consequences, the consequence of a broken cryptographic primitive regarding the past should be considered. Ideally, breaking a primitive should have no impact regarding security and privacy on tag queries in the past. The required property is called *forward security* or *forward secrecy*.

Figure 4.22 shows an example of forward secrecy when using identifier modification. Once a cryptographic primitive is broken and the attacker reveals the inner secret of the tag, he shall not be able to derive earlier tag identifiers, combine this information with past reading logs and thus obtain information regarding past activities of the person carrying the tag.

PERFORMANCE: An obvious evaluation criterion is performance. RFID systems shall return the data regarding a tag within a short time. Besides design and implementation of the backend infrastructure, the speed of reading of tags is very important here.

As explained in section 2.1.4, the speed of reading depends on a number of properties. Regarding protocol design, the amount of data to be transmitted, the number of messages to be exchanged, and the time required for performing calculations are the most relevant properties.

As the data rate of the physical channel is limited, the amount of data to be transmitted over this channel has direct impact on the speed of reading. The number of messages to be exchanged has an even greater impact: It makes a huge difference whether data can be transmitted in a single message or, for example, some data is transmitted in one message, a calculation is performed, the result is then sent back, and an acknowledgement message is transmitted. Thus, roundtrips of messages should be avoided. This is a matter of protocol design.

Backend processing and processing in the tag should also be kept as low as possible. Especially the processing of cryptographic primitives in a tag can become a limiting factor for overall performance. The kind of cryptographic primitives used and the number of invocations of these cryptographic primitives are thus important aspects here. But besides the number of invocations, parallelism is of relevance: It makes a huge difference whether a tag performs a calculation and the reader needs to wait for the answer or whether the calculation can be performed in the background while the tag gets power from a reader. In the following, the possibility of background computations will be denoted by "offline computation" and "precalculation".

As stated in the considerations regarding cryptographic primitives in section 2.4.5, there is a trade-off between resource consumption and speed. How this trade-off is performed is an implementation optimization which is not relevant for the evaluation of the architecture of an RFID system.

Regarding performance, the possibility for caching data is an important aspect. Caching can save resources by decreasing required communication or computation. Further, time is saved if requests can be answered locally instead of requiring a remote query for data. Instead of simple caching, delegation can be performed to enable an entity to perform operations on behalf of another. With delegation, which was introduced to RFID systems in [MSW05], operations can often be performed locally instead of remotely. This can save time and resources.

HANDLING/PRACTICABILITY: A good RFID system should not require a user to pay attention to it for everyday tasks. How far a system adheres to this requirement is also an evaluation criterion.

For instance, alternative channels like written texts (see subsection 4.6.3) often put a burden on the user. In the text scenario, the user is required to enter this text somewhere for certain operations. Another example is to require the use of a physical contact channel instead of the convenient wireless channel.

If such an action needs to be taken for a security sensitive operation, this burden is necessary and justified, but for other operations it should be avoided requiring the user to perform actions. An example can be found in subsection 4.6.1 with the wake-up code printed on the sales slip: Requiring the action to enter the wake-up code somewhere is a viable option as long as the number of items returned is low and a wake-up is not required for other purposes. In other cases, more automatism should be provided to disburden the users.

Another example is that users should not be required to be delayed in their actions until an operation has completed. Time consuming operations should be able to be performed in the background without the user remaining with a tag in the read range of a reader. Even worse, an interrupted operation should not result in an intricate recovery task that requires user involvement.

UNIVERSALITY/SCOPE: As already explained, to avoid tracking by constellation based on traffic analysis, tags shall behave as identical as possible. This can only be achieved by having an RFID system that is application independent. This means that the RFID system needs to be very generic, but it also requires the ability to add application specific extensions without affecting security and privacy in a negative manner.

The scope of RFID systems can range from local, single organizational, closed systems to global, inter-organizational, open systems. For which scope an RFID system architecture is suited is thus an interesting evaluation criterion.

In the previous discussions within this book, only systems of small scope have been considered. This corresponds to the scope that is usually found in literature. The only exception is the mentioned global EPC network. In chapter 6, it will be argued that it makes sense to consider RFID systems in a wider context to gain the potential to operate RFID systems inter-organizationally in a global scope. Therefore, different roles within the RFID systems will be split into new entities to better adhere to the practical requirements.

4.8 Hash-based ID Variation

In this section, a complete protocol that uses hash functions as its only cryptographic primitive is presented. It implements all the main tasks that have been identified and discussed in the previous sections: identification, authentication, and modification. The design had security and privacy in mind so that the protocol is safe from eavesdropping and cannot be troubled by spoofing or replay attacks. Location privacy is enhanced by changing the tag identifier on every read attempt in a secure manner.

The approach has originally been published as *Hash-based Privacy* in the proceedings of two major pervasive computing conferences (see [HM04b] and [HM04a]) in 2004. It has been cited in many publications like the German BSI study regarding risks and prospects of the use of RFID systems [BSI05], the Italian guidelines for the employment of the RFID systems in the public administration [CNIPA07], in the current study [BMBF07] regarding technology-integrated data security in RFID systems, in a number of publications regarding security and privacy of RFID systems [Avo07], and appeared even on slides of a lecture. Usually it is referred to the approach as the "approach/protocol/procedure/... of Henrici and Mueller" or "Hash-based ID variation", therefore the heading of this section.

As the previous subsections have shown that the topic is sophisticated, the approach will be explained in several steps. At first, the basic concepts will be presented, and then the protocol will be explained in detail. The section will close with a security analysis and an evaluation.

4.8.1 Basic Concepts

The basic idea of the approach follows the presented tasks: The tag identifier is used for identification. Authentication is used against tag cloning and mimicking and for securing the protocol. Modification, i.e. changing the tag identifier, is used to counteract unwanted recognition and tracking. An overview of the basic tag organization in the approach is depicted in figure 4.23.

SENDING UPDATE INSTRUCTIONS: The basic organization is the same as shown in figure 4.21 in subsection 4.5.3. The internal tag identifier is revealed to the outside in hashed form as tag identifier *ExtID*. Tag identifier modification is done by transmitting a *ChangeInfo* instead of sending a new identifier in clear. This removes the incentive for eavesdropping. A new identifier is calculated using the former identifier and the *ChangeInfo* by calculating $ID^* = ID \oplus ChangeInfo$ in which $\oplus$ is an xor-operator and the domain of the identifiers and the *ChangeInfo* are the same.

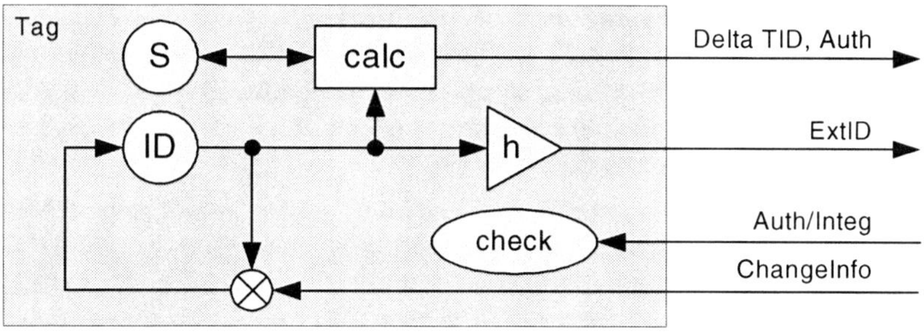

Fig. 4.23. Simplified overview of tag organization in the hash-based ID variation approach

ENSURING INTEGRITY AND MUTUAL AUTHENTICATION: Hash functions are used as primitive for ensuring integrity of messages and for performing mutual authentication. In a first message from tag to legitimate entity, the tag identifies itself using *ExtID* and proves its identity using a hash value. In a second message from the legitimate entity to the tag, the legitimate entity proves its identity and secures the integrity of the *ChangeInfo* using another hash value. The mutual authentication is required to counteract spoofing and mimicking attacks. Furthermore, the approach requires tag and legitimate entity to have a common state so that any state changes need to be performed in coordination of both peers.

USING TRANSACTION IDENTIFIERS: The hash function that generates the tag identifier *ExtID* takes the state information S as an additional preimage. This state information includes a transaction identifier that will be called *TID* in the following. By including such a transaction identifier *TID*, it is ensured that no replay attacks can be performed. The transaction identifier *TID* changes on every tag query by simply incrementing a counter.

USING THE DIFFERENCE OF TWO TRANSACTION IDENTIFIERS: As the hash value that is used for tag authentication changes on every query due to a changing transaction identifier *TID*, the legitimate entity needs to know which *TID* the tag currently uses. If the tag is only queried by the legitimate entity without errors, the legitimate entity can calculate the current *TID* by incrementing a counter. But if the tag is queried by other entities or a communication error occurs, the legitimate entity does not know how many times it needs to increment the transaction identifier counter to close on. This is again the synchronization problem that has already been highlighted in subsection 4.5.3: Trying all values until one fits is no solution since this could be used as a means for a denial-of-service attack. Including the current transaction identifier in the message is no solution because this identifier could be used by an attacker for unwanted recognition and tracking.

The idea for solving the problem is to transmit the difference of the current transaction identifier *TID* and the transaction identifier that was used in the last successfully completed transaction. The latter will be denoted by *LST* in the following. This way, the transmitted value is usually one, i.e. $\Delta tid = 1$, because initially $tid = lst$, the transaction identifier is incremented $tid^* = tid + 1$, and the difference between the current transaction identifier and the identifier of the last successfully completed transaction $\Delta tid = tid^* - lst = 1$ is transmitted. Only in cases of errors and attacks other differences occur. But a successful message exchange will immediately reestablish synced transaction identifiers so that the transmitted value Δtid cannot be abused for unwanted recognition and tracking.

COPING WITH MESSAGE LOSS: The protocol needs to be resilient against the loss of protocol messages. Therefore, the communication peers need to consider that a sent message might not reach its receiver. This becomes a problem if a message is used to trigger a state change in a peer. The protocol uses two messages. If the first message reaches the legitimate entity, it triggers a state change there. The legitimate entity handles a possible loss of that first message. The following second message from legitimate entity to tag triggers a corresponding state change in the tag. To cope with loss of this second message, for each tag two records are kept in the tag database of the legitimate entity: One record that still stores the old state information and another record that stores the new state information. The two records point to each other using a field *PTR* that contains the index field of the associated record.

4.8.2 Protocol Realization

In the previous subsection, the concepts that are employed in the approach have been introduced. In the following, the approach is presented in more detail. The data organization of tag and legitimate entity is shown and the message exchange between the two communicating peers, i.e. tag and legitimate entity, is explained.

In the introduction of the basic concepts in the previous subsection, a number of values have been mentioned that need to be stored within the RFID tags. Thus, each tag needs to have changeable storage for the following variables:

- current internal tag identifier *ID*,
- current transaction identifier *TID*, and
- transaction identifier of the last successfully completed transaction *LST*.

The legitimate entity that is responsible for the tag needs to have a database table in which information about the tags is stored. Much of this data corresponds to the one in the tags because legitimate entity and tags remain in synchronization. The database table of the legitimate entity needs to have the following fields:

- hash value of the current internal tag identifier, i.e. *ExtID*, acting as primary index of the table,
- current internal tag identifier *ID*,
- transaction identifier of the last transaction *TID*,
- transaction identifier of the last successfully completed transaction *LST*,
- pointer to associated table row *PTR*, and
- reference to tag data / user data *DATA*.

The fields of the tag are initialized to the following values: The internal identifier *ID* is set to a random value *id*. Another random number *tid* is stored as transaction identifier *TID*, and the same number $lst = tid$ is stored in the field for the last successfully completed transaction *LST*.

A corresponding row in the database table of the legitimate entity must be created. The internal tag identifier field *ID* is set to the internal identifier *id* of the tag, the primary index *ExtID* is set to the hash value of the internal tag identifier $extid = h(id)$. The fields *TID* and *LST* get the corresponding value $tid = lst$ of the tag. The pointer *PTR* is not set since no associated table row initially exists in the database.

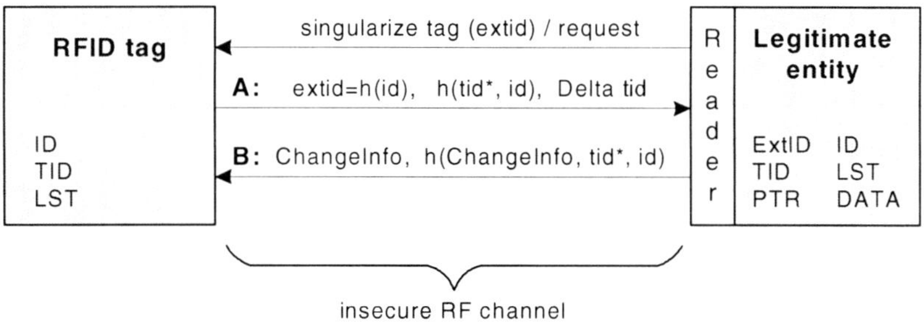

Fig. 4.24. Message exchange in the hash-based ID variation protocol

The regular operation of the protocol proceeds as follows: A tag is singularized out of many by any standard method like binary tree walking or any other anticollision protocol (see subsection 2.1.4). In this stage of operation, a tag exposes no other information than the hash value of its internal tag identifier *ID*, namely $extid = h(id)$. This hash value *extid* is used for addressing and identifying the tag.

When queried, the tag increments its transaction identifier *TID* by one so that it obtains $tid^* = tid + 1$ and sends the data that is depicted as message *A* in figure 4.24 to the reader: *extid* (if not already known to the reader out of the operation of the anticollision protocol), the hash value $h(tid^*, ID)$, and the difference between the tag's current transaction identifier tid^* and the identifier of the last successfully completed transaction $\Delta tid = tid^* - lst$. The reader then forwards the received data to the legitimate entity.

In this message, the tag identifier *extid* identifies the tag in the database of the legitimate entity. The hash value $h(tid^*, id)$ has the purpose of counteracting replay attacks: It changes at every read attempt due to the changing transaction identifier. The legitimate entity checks whether this value is correct. As the internal tag identifier and the current transaction identifier is only known to the two communication endpoints, this way the identity of the tag is proven and tag mimicking or even cloning thus not possible. Including the internal tag identifier as preimage is only mandatory if the domain of the transaction identifiers is that small that an attacker can perform a brute force or dictionary attack. The difference Δtid of the transaction identifier values tid^* and *lst* is used to enable the legitimate entity to calculate the current transaction identifier used by the tag. Since it is only a difference with a value of 1 if no special event, i.e. loss or change of a message or a tag query by an illegitimate entity, has occurred, no information that could be utilized by an attacker for unwanted recognition and tracking is revealed.

The legitimate entity that receives message A from the reader selects the database record of the tag with the received *extid* as key. If such a record is not found, the tag is not registered to that entity. The stored last successfully completed transaction identifier *lst* of the tag and the received Δtid are summed-up so that the legitimate entity obtains the current transaction identifier tid^* of the tag. Now the hash $h(tid^*, id)$ is calculated based on the stored values. If the value does not match the one in the message, the message is discarded. If the message proves to be valid so far, the calculated tid^* and the stored *tid* are compared. If the calculated tid^* is not higher than the stored *tid*, a replay attack is in progress and the message is discarded. If everything is fine, the calculated tid^* is stored in the *TID* field of the record row of the database table and the message is processed further.

Now a random number *ChangeInfo* that should have the same number of bits as the internal tag identifier *id* is created. With this value *ChangeInfo*, a new internal tag identifier id^* is created by performing $id^* = ChangeInfo \oplus id$.

ExtID	*ID*	*TID*	*LST*	*PTR*	*DATA*
$extid = h(id)$	id	tid^*	lst	$h(id^*)$	$data$
$extid^* = h(id^*)$	id^*	tid^*	$lst^* = tid^*$	$h(id)$	$data$

Table 4.1. Records regarding a tag after processing message A

If an associated table row that is referenced by the pointer *PTR* already exists, the internal identifier field *ID* of this record is updated to the new internal identifier id^*, and its key *ExtID* is updated to the hash value $extid^* = h(id^*)$. Otherwise, a new row is appended to the database table by inserting the new internal identifier id^* as *ID* and $extid^* = h(id^*)$ as *ExtID*. The reference to the tag data *DATA* is cloned and the *PTR* field of the new row is set to the *ExtID* of the current row. The *PTR* field of the current row is updated to the newly calculated key $extid = h(id^*)$ so that now the pointers *PTR* point to the respective other row. The transaction identifier field *TID* of the newly selected row is updated to the tid^* value, its last successfully completed transaction identifier *LST* gets the same value. The two database rows of the tag after processing message A are shown in table 4.1. The legitimate entity can use the reference *data* to access the data that is associated to the tag.

Now a reply message B, see 4.24, containing *ChangeInfo* and a hash value $h(ChangeInfo, tid^*, id)$ is created and sent back to the tag. The tag checks the hash value using the received *ChangeInfo* and the current values of the internal identifier *ID* and the transaction identifier *TID* stored in the tag. If it is not correct, the message is discarded and no further action is taken. Otherwise, the tag updates its internal tag identifier *ID* to the value $id^* = ChangeInfo \oplus id$ and sets its last successfully completed transaction identifier *LST* to the value of the transaction identifier *TID*. Now the tag has a new internal identifier *ID* whose hash value $extid^* = h(id^*)$ will be used as tag identifier at the next read attempt.

Note that except for the time before the first update, there are always two table rows for each tag in the database. The idea is that the next tag query after a successful message exchange acknowledges the success of that message exchange. Always, the row for the last acknowledged message exchange and the row from the previously tried message exchange are stored in the database. This way, one of the two rows is valid for the given tag at any time.

By having the two rows for each tag in the database, communication errors or purposeful errors due to attacks by an attacker up to the strength of *Mallory* (see figure 3.2) cannot bring legitimate entity and tag out-of-sync.

As in all protocols for RFID tags, special care needs to be taken regarding sudden power interruption. In any case, the tag needs to go into a valid state when such an event occurs. In this protocol, it is essential that the change of the internal tag identifier and the update of the last successfully completed transaction *LST* are performed as an atomic operation: either both values or updated or none. Otherwise, legitimate entity and tag could get out-of-sync which would render the tag unidentifiable.

4.8.3 Security Analysis

In this subsection, the security characteristics of the presented approach will be analyzed. After a repetition of the goals to be reached, at first some basic security and privacy considerations are made. Afterwards, a more detailed discussion regarding common attacks like the replay of messages will be performed. Within these discussions, examples are given that also provide a better understanding of the protocol operation.

Goals to be reached

The goal of the protocol is to implement the functionality that has been identified in section 4.3. This way, the high-level goals, i.e. maintaining data security, preventing counterfeiting, preventing illegitimate access, preventing unwanted recognition and tracking, and coping with denial-of-service can be met in an optimal manner.

The functionality shall be implemented in such a way that protection against an attacker like *Mallory* (see section 3.5 on attacker classification) on protocol level is obtained. This means that the protocol needs to withstand active attacks like message interception, message modification, and replay attacks. Normal operation shall also be provided in the case of errors that are not caused by an attacker: Reliability against errors like message loss is seen as a basic requirement regarding the functionality.

Basic security and privacy considerations

In the following, some high-level considerations regarding security and privacy of the protocol are made. At first, the security measures and afterwards the resulting characteristics are presented before this paragraph concludes with the limitations of the protocol.

SECURITY MEASURES: The two messages used in the protocol are authenticated using a hash function that has preimages only known to tag and legitimate entity. Therewith, the validity of the origin of the messages is proven since an attacker does not have the information that is necessary for creating valid hash values.

Of course, an attacker can overhear valid messages and replay them at a later time. Transaction identifiers are introduced to counteract such activities to succeed: Messages that are based on an old transaction identifier are discarded by tag and legitimate entity so that no harm can be done by old messages.

If data besides the hash values is transmitted, its integrity is ensured by including that data into the hash calculation. This is the case for the *ChangeInfo* in message *B*. If the message is altered, checking the hash value will lead to an invalid result so that the message will thus be discarded by the receiver.

By a complete message exchange, the internal tag identifier gets changed. This way, only the legitimate backend entity is able to identify the tag because an attacker does not have the required information to link independent queries so that the requirement of indistinguishability is fulfilled.

RESULT: Tag cloning is effectively prevented by keeping an inner tag state that is never revealed to the outside world. An attacker like *Mallory* does not learn information that could be used for deriving the inner tag state because the inner tag state is only used as preimage in hash calculations, and the identifier update messages are valueless without knowing the current inner state.

Mimicking a tag is possible in some circumstances: An attacker could act as a reader and record the tag's answers to the queries. Later, the attacker could replay the answers in the same sequence to the queries of a legitimate reader. This way, the tag still appears to be in place albeit already taken away.

But such an activity can be detected and even effectively be prevented. Multiple queries that do not result in a change of the internal tag identifier result in Δtid values greater than one. Suspiciously high values indicate that an attack is in progress.

If it shall be ensured that such a mimicking is not performed, the legitimate entity needs to check for a successful message exchange. Such a successful message exchange is given if the next tag query indicates that the internal tag identifier has successfully been modified. Thus, a second tag query can be used as an acknowledgement for the previous transaction.

Location privacy is gained by the regular modification of the tag identifier. This way, the tag is only recognizable and traceable by the legitimate backend entity. The criterion of indistinguishability that has been shown in figure 4.1 in section 4.2 is met because tag identifiers are completely independent of each other.

Only hash values and the Δtid are revealed to the outside. The hash values appear as random oracle and do thus not provide exploitable information to an attacker. The Δtid is usually equal to one so that it does not provide exploitable information either. As stated in section 4.8.1, using the difference of transaction identifiers solves the problem of recognition and tracking based on revealed transaction identifiers. Of course, it needs to be ensured that a successful modification of the internal tag identifier is the usual case, i.e. all readers except those operated by attackers should try to complete the protocol successfully. Otherwise, the *Δtid* values could grow and could be used for unwanted recognition and tracking – at least by constellation.

By keeping two records in the database table, message loss – whether by coincidence or caused by an attacker – does not bring tag and legitimate entity out-of-sync. The tag can still be correctly identified after message loss since the record with that old tag identifier is still present in the database table.

Message loss can be detected afterwards by the legitimate entity on the basis of a *Δtid* value that is unequal to one. Suspiciously high values attract attention and indicate an ongoing attack so that counteractive measures can be taken.

The internal tag identifier and the transaction identifiers have limited validity: All these variables change with each completed protocol run. This lowers the potential impact of successful attacks.

In the worst case in which the internal state information of the tag is revealed, e.g. by physical extraction of the attacker *Phyllis*, the attacker can only obtain previous tag identifiers if he has overheard all the *B*-messages in between. Thus, the protocol provides limited forward secrecy.

In the stated worst case, the attacker becomes able to imitate or clone the tag or bring the tag and the responsible legitimate entity out-of-sync which results in a permanent denial-of-service. But the internal state information is only valid and usable until two successive protocol runs that have not been overheard by an attacker have been successfully completed.

LIMITATIONS: If the reply message *B* of the legitimate entity does not reach the tag, the tag identifier is not changed. This can happen due to loss, interception, or blocking of that reply message. As a result, the tag will use its current identifier again in the next tag query and the *Δtid* value increases.

Avoine (et. al.) stated two attacks regarding this behavior in [AO05a] and repeats the descriptions in [Avo05b] as well as in [Avo05a]. On the one hand, he states an attack he calls "Attack Based on Non-Random Information". The idea is to query a tag very often without sending a reply message. This leads to unusual high *Δtid* values. Under the assumption that the tag is not successfully queried by a reader that completes the protocol, the attacker can query the tag again later and recognize it by the high *Δtid* value.

This attack is regarded uninteresting in practice as long as completing the protocol is the standard procedure by each reader that is not operated by an attacker. Instead of watching for special *Δtid* values, the attacker could also simply use the tag identifier $h(id)$ for identification because it does not change if no reply message reaches the tag.

The second stated attack by Avoine is called "Attack Based on Refreshment Avoidance". The idea here is that an attacker prevents the reply messages from reaching the tag so that the tag identifier is not changed. For realization, Avoine proposes to query the tag again before the reply message of a legitimate query reaches the tag.

The author of this book considers this attack to be impractical. If an attacker has the capability to bring a device that near to a tag so that it can perform such an action when required, he could also use this device directly for making the object recognizable. For instance, this device could emit a special signature when queried by the attacker. Instead of an active device that performs the tag query attempts, a passive RFID tag would be sufficient for an attacker. Ideally, this second tag introduced by the attacker should only respond to queries that are performed by an attacker so that it cannot be detected easily.

Although these two attacks are impractical, Avoine also showed nifty attacks that can lead to a permanent denial-of-service if the protocol is not properly implemented. These attacks will be discussed later.

Overview of protocol behavior in abnormal cases

An analysis of susceptibility to lost packets, packet interception, and replay attacks is performed in this paragraph. For the purpose of a comprehensible description, the discussion is based on an example.

Tag				*Legitimate entity*			
ID	*TID*	*LST*	*h*(*ID*)	*ID*	*TID*	*LST*	*PTR*
17	2	2	*h*(123)	123	2	1	*h*(17)
			h(17)	17	2	2	*h*(123)

Table 4.2. Valid synced state of tag and legitimate entity as starting basis

Table 4.2 shows a setup that is the basis for the following considerations. The tag has a current internal identifier 17. The transaction identifier *TID* and the identifier of the last successfully completed transaction *LST* have the same value, here 2. This indicates that the last message exchange has been completed successfully so that the internal tag identifier has been updated. The state of the legitimate entity is synced to the one in the tag. In addition, a second record containing the internal tag identifier that was used previously is present. The value of the transaction identifier *TID* is the same in both records since it has been updated in both records upon reception of an authenticated message from the tag. The two records point to each other using the *PTR* field.

Tag				*Legitimate entity*			
ID	*TID*	*LST*	*h*(*ID*)	*ID*	*TID*	*LST*	*PTR*
854	3	3	*h*(854)	854	3	3	*h*(17)
			h(17)	17	3	2	*h*(854)

Table 4.3. State after regular update

REGULAR TAG QUERY: If another tag query with a complete message exchange is performed successfully, the state looks as shown in table 4.3. The internal tag identifier was changed from 17 to 854 in this example. The situation is similar to the previously described one: The only differences are that all transaction identifiers were incremented by one and that the old record with $h(123)$ as key was updated to a new one.

Such a regular update renders all messages that were sent between tag and legitimate entity invalid. Messages of previous tag queries as well as the message of the current tag query are not based on a transaction identifier that is newer than the ones currently stored in the tag or in the database of the legitimate entity. Thus, replay of any of these messages in arbitrary direction is useless for an attacker. As the transaction identifiers are old, the messages will be immediately discarded by the respective receiver.

	Tag				*Legitimate entity*		
ID	*TID*	*LST*	*h(ID)*	*ID*	*TID*	*LST*	*PTR*
17	3	2	$h(123)$	123	2	1	$h(17)$
			$h(17)$	17	2	2	$h(123)$

Table 4.4. State after one loss of message *A*

REQUEST MESSAGE *A* LOST OR INTERCEPTED: One of the abnormal cases is message loss. The protocol uses two messages, *A* and *B*, see figure 4.24. Table 4.4 shows the state after loss of one message *A*. The following are the three different cases that can occur after such a loss of the first message *A*:

- Packet never inserted again.
- Attacker inserts message again before other communication between tag and legitimate entity occurred.
- Attacker inserts message again after other communication between tag and legitimate entity occurred.

If the message is lost or intercepted and is never inserted again, the legitimate entity will get no information about the tag query and will thus not perform any changes in the tag database. In the next tag query, the tag will use a new transaction identifier and the Δtid value increases. Thus, the legitimate entity is able to obtain the current transaction identifier used by the tag and is therewith able to check the authenticity of that new message. Message loss does thus not bring tag and legitimate entity out-of-sync so that the protocol is reliable against a loss of message *A*.

If the message is lost or intercepted and inserted again before other communication between tag and legitimate entity occurred, the result is a simple delay of the message. This has no negative effect on the system. It only increases the possibility that the reply message does not reach the tag any more because the tag might have been moved out of range of the reader before.

The most interesting case is the third one in which a message is lost or intercepted and inserted again after other communication between tag and legitimate entity has occurred. This case has two subcases: If another message of the tag reaches the legitimate entity, the intercepted message is rendered useless since it would be discarded by the legitimate entity due to too small a transaction identifier. In contrast, as long as no other message of the tag reaches the legitimate entity, the intercepted message remains valid for the legitimate entity. Thus, the attacker can collect many request messages sent by the tag. If one of these later messages is forwarded to the database, the first intercepted message is rendered useless due to too small a transaction identifier. If the first intercepted message is forwarded to the database, a normal update occurs there, but the reply is discarded by the tag due to a non-current and thus invalid transaction identifier. Later communication between tag and database is still possible employing the unchanged database entry so that tag and legitimate entity do not get out-of-sync.

	Tag			*Legitimate entity*			
ID	*TID*	*LST*	$h(ID)$	*ID*	*TID*	*LST*	*PTR*
17	3	2	$h(854)$	854	3	3	$h(17)$
			$h(17)$	17	3	2	$h(854)$

Table 4.5. State after one loss of reply message B

REPLY MESSAGE B LOST OR INTERCEPTED: Another abnormal case is the one, in which the reply message B does not reach the intended receiver, i.e. the tag, as planned. Here, one can distinguish the same cases as in the case of a loss of message A:

- Packet never inserted again.
- Attacker inserts message again before other communication between tag and legitimate entity occurred.
- Attacker inserts message again after other communication between tag and legitimate entity occurred.

If the message B is lost or intercepted and never inserted again, the legitimate entity has updated its database upon reception of the message A, but the response message does never reach the tag. Thus, the tag does not perform an update. The situation is shown in table 4.5: The tag still has the old internal tag identifier, and the last successfully completed transaction variable *LST* still has the old value.

But future communication between tag and legitimate entity and the identification of the tag is still possible employing the unchanged database entry, here: $extid = h(17)$. Thus, no harm is done: The old database record is not overwritten until the other record has been addressed by the tag proving that that other entry is currently valid.

If the message is inserted again into the communication channel before other communication between tag and legitimate entity occurred, this results in a delay of the message. This has no negative effect; it only increases the possibility that the reply message does not reach the tag. The effect would be a lasting loss of the message thus resulting in the previously described subcase.

If the message is inserted again after other communication between tag and legitimate entity occurred, the tag will discard the message because the transaction identifier that is used in the message will be non-current and thus invalid.

SUMMARY: The previous considerations showed what happens in the case of message loss, message interception, and message replay. The discussion highlighted how the protocol deals with such circumstances. The result is that the protocol works reliably and that it cannot be harmed by packet interception or replay attacks.

Security pitfalls

In the initial publication of the protocol in [HM04a], it has been assumed that the use of the xor-operator is adequate to conjugate the preimages of the hash functions, i.e. $h(p_1, p_2) := h(p_1 \oplus p_2)$. Avoine explained in [Avo05a] that this assumption was wrong.

In previous publications, i.e. [AO05a] and later again in [Avo05b] in which weaknesses of several protocols have been highlighted, Avoine presented a first weakness caused by the xor-operator. If the *ChangeInfo* is zero, it is the neutral element of the xor-operator. In this case, the hash values in message *A* and message *B* are the same. An attacker can thus bring tag and legitimate entity out-of-sync by querying the tag and answering with faked *B* messages in which the *ChangeInfo* is zero and the hash value eavesdropped from the message *A* is used. This will update the last successfully completed transaction identifier *LST* of the tag whereas the one in the database of the legitimate entity remains the old one. The result is that the tag can no longer be identified by the legitimate entity and that the tag identifier does not change any more so that the tag becomes trackable by the attacker. Fortunately, the attack can easily be avoided if the tag does not accept a *ChangeInfo* that is zero.

In [Avo05a], Avoine presents an interesting attack that is based on the xor-operator and the known increase of the transaction identifier in each transaction. Based on the concept used in the attack presented in that paper, the following proceeding of an attacker is possible: The attacker overhears the reply message *B* of a tag query and thus obtains $h(ChangeInfo \oplus tid^* \oplus id)$ as shown in figure 4.24 but with the xor-operator used for conjugating the preimages. In the next tag query, the tag sends the hash value $h(id^*)$ of the new tag identifier id^*, $h((tid^* + 1) \oplus id^*)$, and $\Delta tid = 1$. The legitimate backend entity would now reply with $ChangeInfo^*$ and the hash value $h(ChangeInfo^* \oplus (tid^* + 1) \oplus id^*)$. Instead, the attacker sends $ChangeInfo^* = 1$ and the overheard $h(ChangeInfo \oplus tid^* \oplus id)$ to the tag. With a probability of $\frac{1}{2}$, the faked message is accepted by the tag so that tag and legitimate entity are brought out-of-sync.

The attack is based on the equivalence of $1 \oplus (tid^* + 1) = tid^*$. This equation is true if the last digit of tid^* is zero. This is the case with the probability of $\frac{1}{2}$ if the transaction identifier is implemented as a counter.

For the attack to succeed, the hash value $h(1 \oplus (tid^* + 1) \oplus id^*)$ expected by the tag for $ChangeInfo^* = 1$ must equal the overheard hash value $h(ChangeInfo \oplus tid^* \oplus id)$.

This is true if the presented equivalence $1 \oplus (tid^* + 1) = tid^*$ holds which will be shown in the following: With the presented equivalence $1 \oplus (tid^* + 1) = tid^*$, the expected hash value $h(1 \oplus (tid^* + 1) \oplus id^*)$ can be transformed to $h(tid^* \oplus id^*)$. By inserting the identifier update rule $id^* = ChangeInfo \oplus id$, the latter can be transformed to $h(tid^* \oplus (ChangeInfo \oplus id)) = h(ChangeInfo \oplus tid^* \oplus id)$ which equals the overheard hash value.

The consequence is that the xor-operator is a bad choice for conjugating the preimages of the hash functions. To avoid the stated weaknesses, a more appropriate compression function needs to be used.

In [LHLL05], the authors claim to have found a simpler protocol that is based on the one presented in this section. They removed the transaction identifiers *TID*. Instead, the reader sends a random number to the tag that replaces the transaction identifier in the hash operations and also acts as *ChangeInfo*.

The removal of the transaction identifiers makes the protocol susceptible to replay attacks that cause tag and legitimate entity to get out-of-sync. This can be done by intercepting a reply message from the legitimate entity. Then the tag is queried again with another random number and the reply message is intercepted and discarded. The legitimate entity thus assumes that the first reply message has been lost and overwrites the database record. In the third step, the tag is queried again with the random number used in the first step. The message from the tag is intercepted and discarded so that it does not reach the legitimate entity. The message that has been intercepted in the first step is inserted as reply message to the tag into the communication. This message triggers an identifier modification in the tag so that tag and legitimate entity are out-of-sync now.

Both examples showed that designing a secure protocol is difficult. For the protocol concept of "Hash-based ID variation", no flaws have been found yet. Thus, if the protocol is implemented properly using an appropriate compression function it can be regarded secure.

4.8.4 Variants

A number of variants of the presented protocol are imaginable. In the following, one of them shall be presented. In this variant, no *ChangeInfo* is transmitted any more. Instead, the tag calculates the new internal identifier using a hash function $id^* = h_f(id)$ upon reception of the correct authentication message B. The identifier modification is therewith based on the scheme shown in figure 4.17. Modification schemes that calculate a new identifier as the hash value of an old one are used in a number of proposed protocols, e.g. in the concept presented in [OSK03] or the flawed protocol presented in [LAK06]. The hash function h_f used for identifier update must of course be different from the one used for calculating the external tag identifier $extid = h_g(id)$ since otherwise the new internal identifier and the current external tag identifier would be the same, i.e. $extid = id^*$ which would break protocol security.

When there is no *ChangeInfo* any more in message B, the hash functions in message A and message B have the same preimages. To avoid this, one needs to perform the hash calculations with an additional preimage that makes the hash values different or needs to use different initialization vectors in the hash calculations. Alternatively, the hash value can be split into two parts if the number of bits is high enough. Such a split has been used in the protocol presented in [LHLL05]. One half of the hash value is then used for authentication in message *A* and the other half is used for authentication in message *B*.

The advantage of the presented variant that performs a hash calculation for the update of the internal identifier instead of using the *ChangeInfo* is twofold. First, the amount of data to be transmitted decreases since no *ChangeInfo* needs to be transmitted. Second, the variant provides better forward secrecy. Using the *ChangeInfo*, the forward secrecy was limited because with a current internal identifier and the *ChangeInfo*-values, the previous internal identifiers could be calculated. The hash operation in the variant is completely one-way so that it is not possible to derive a previous tag identifier from a current internal tag identifier even if the communication has been overheard. The disadvantage is that two hash functions are required (or need to be emulated using a single one by using an additional preimage or by using different initialization vectors) and that an additional hash operation is required. This is again a trade-off between security and performance.

Note that the variant of the *Hash-based ID variation* scheme presented in this section can be regarded as an advanced extension to the self-contained modification scheme presented in [OSK03]. In that scheme, there are also two hash functions, one for calculating the external tag identifier and one for updating the internal tag identifier. In contrast to that scheme, the additional transaction identifier concept results in a scalable solution that is not susceptible to denial-of-service attacks. An even more advanced extension will be presented later in section 7.1.1.

4.8.5 Evaluation

In this subsection, the protocol is evaluated according to the criteria that have been presented in section 4.7. This evaluation currently stands alone as there are no other protocols for reference. But a comparison with other schemes will be presented in chapters 6 and 7 in which other protocols and concepts are introduced in detail.

SECURITY: As shown in the security analysis in subsection 4.8.3, the protocol is tolerant regarding message loss, spoofing, and replay attacks. As long as the identifier modification takes place on every usual tag query, i.e. each query that is not performed by an attacker, the scheme provides location privacy and thus prevents unwanted recognition and tracking of the tag.

As a modification on every usual tag query requires a central legitimate entity, this entity needs to be trusted by all organizations and users drawing on its service. This is a huge disadvantage so that a central authority should be avoided. Thus, in the next chapter, the scheme will be extended to support a distributed organization.

RESOURCES: The tag requires non-volatile storage for three variables, and the database needs to store ten values for each tag (five for each of the two rows). Regarding computation, each of the two peers needs to perform three hash operations for a complete transaction. From tag to the reader, three values need to be sent, in the opposite direction only two.

SCALABILITY: As stated in the security evaluation, the identifier modification needs to take place on every tag query. Thus, a central legitimate entity is required so that each reader can forward the message received from the tag to it. Such a central entity is not desirable from a scalability point of view. A decentralized solution that enables load sharing among many machines at different locations is preferable. In the next chapter, the solution will be extended to support such a structure, but the current protocol does not have these possibilities.

SUSTAINABILITY: The described protocol uses a particular hash function so that no migration path to other hash functions is present. Nevertheless, the scheme can easily be extended to support such an option. One simply needs to add an additional field in the database table of the legitimate entity that denotes the hash function that is used in the hash calculations for the particular tag. This way, the legitimate entity can cope with tags using different hash functions.

The impact of a broken cryptographic primitive or a successful physical attack by *Phyllis* is limited to a single tag as there are no shared secrets in the scheme. But for the respective tag, the attacker gains complete control over the tag: He can perform a permanent denial-of-service and gets able to perform tag cloning.

Forward secrecy has already been discussed in subsection 4.8.4. It can be regarded to be present already in the basic protocol. With the variant presented in subsection 4.8.4, forward secrecy even becomes perfect.

PERFORMANCE: A complete transaction requires two messages, i.e. one roundtrip, between tag and legitimate entity. For mutual authentication, this is the optimal result. The tag can precalculate the first two hash operations, i.e. $extid = h(id)$ and its authentication hash value. The tag just needs power from a reader to do that. The third hash calculation that checks the integrity of the content of message B and that authenticates the legitimate entity needs to be performed after reception of message B since the *ChangeInfo* is not known to the tag before. But the tag just needs power to do that; no action of the reader or the legitimate entity is required any more. These are very good characteristics.

If multiple tag queries are performed by the same reader, nevertheless the complete protocol needs to be run including the requirement to contact the legitimate entity. In applications in which lots of tag reads are performed, e.g. for regularly checking whether tags are still at the same place, this results in lots of network traffic to the legitimate entity and much load there. The possibility to recognize a tag by the same reader without requiring to contact the legitimate entity again would be useful because this would make caching of data possible and reduce overall resource consumption. But such an optimization is not possible using the presented protocol.

HANDLING/PRACTICABILITY: The communication uses only the convenient wireless channel so that no explicit user action is required. A disadvantage is that the identifier modification cannot be completed if the tag is moved out of the field of the reader too early. Depending on the time required for performing the required operations in a real implementation, this might affect location privacy negatively.

UNIVERSALITY/SCOPE: The protocol is application independent and can thus be used in any application scenario. After identification of the tag, the legitimate entity can perform application specific actions with any other entity requesting data associated to the tag.

The protection of location privacy is only ensured if the identifier modification takes place regularly, i.e. ideally on every tag query. Thus, the scheme would need to be implemented within a global, inter-organizational scope. Unfortunately, the scalability issues with the central legitimate entity do not allow this. This makes the protocol an academic proposal so that it is not yet suited for practical application. As already stated, the next step in addressing this issue will be performed in the next chapter.

4.8.6 Hash-based ID Variation Summary

In this section, the "Hash-based ID variation" protocol has been presented. This protocol implements the main tasks, i.e. identification, authentication, and modification, and is a good example of the current state-of-the-art in RFID protocols that focus on provisioning of security and privacy.

At first, the basic concepts used in the protocol have been introduced. The protocol uses identifier update instructions instead of a transmission of a new identifier, ensures message integrity and mutual authentication using hash functions, uses transaction identifiers to counteract replay attacks, and keeps two records per tag in the tag database to cope with message loss. To prevent the abuse of the transaction identifiers for unwanted recognition and tracking, only differences are submitted.

Afterwards, the protocol has been presented in more detail. This includes the use of two protocol messages and the use of the data that is submitted within these messages. The subsequent considerations about security showed not only the characteristics of the protocol but also highlighted the general requirements regarding "good" solutions. After presentation of a variant, the protocol has been evaluated according to the criteria that had been introduced in a preceding section of this chapter. The result is that the protocol has good characteristics but that it can only be used with a central backend entity which prevents practical application.

4.9 Summary

The aim within this chapter was to examine approaches for securing RFID systems. After the identification of a number of goals in the previous chapter, the topic of this chapter was to reach these goals. Like in the literature, the focus was laid on securing the wireless communication between tags and readers in a resource-friendly manner. Further, data security and location privacy have been considered.

In the first section, it has been shown that data should be stored in the backend and not on tags. This way, the data needs not be transmitted over the insecure wireless channel. Further, more fine-grained access control and more flexible data processing can be performed in the backend.

Afterwards, the security and privacy goals which have been introduced in the previous chapter were discussed again with the aim to find entry points for developing solutions for reaching these goals. Based on these considerations, a number of functionalities that need to be implemented in tags was derived. These basic functionalities, i.e. identification, authentication, and modification, have been presented in the subsequent section.

After a section about general considerations regarding the implementation of such functionalities that dealt with the limitations and primitives, the functionalities have been discussed in detail. First, tag identification and backend identification have been distinguished and the methods for performing identification presented. Second, authentication has been considered. Two main classes, i.e. single message authentication and message exchange based authentication, have been distinguished there and their pros and cons presented. Third, identifier modification has been addressed and categorized into self-contained modification and message exchange based modification.

The subsequent section showed additional building blocks that can be employed in the creation of solutions for reaching the goals. It is possible to distinguish different tag states, one can evaluate lower layer information, and one can make use of alternative communication channels in addition to the wireless channel. These additional building blocks and the different possibilities for implementing the main functionalities can be combined in various ways for creating RFID protocols. For evaluating different solutions, evaluation criteria have thus been discussed in the successive section.

In the last section before this summary section, the "Hash-based ID variation" protocol has been presented. It acted as a comprehensive example of the state-of-the-art for integrating the required functionalities, i.e. identification, authentication, and modification, into a single protocol. The protocol has been discussed in detail by first introducing the concepts, afterwards presenting the actual protocol realization and a security analysis. The considerations finished with a presentation of protocol variants and an evaluation of the protocol. The evaluation results attest overall good characteristics to the protocol. But a central backend entity is required for ensuring location privacy. This main limitation that restricts scalability and therewith the scope of the protocol hinders practical application. The next chapter thus deals with a concept for the removal of these limitations.

5

Pseudonymization Infrastructures

The promising "Hash-based ID variation" protocol that has been presented in the previous chapter has a scalability problem: On each tag query, the reader has to contact a central legitimate entity that recognizes the tag and performs the identifier change. A distributed responsibility would solve the problem of requiring a central entity, but the tag identifier does not contain information which entity is responsible because such information could be abused for unwanted recognition and tracking by an attacker. Instead, the tag identifier needs to be unstructured because each reader is a potential attacker. This chapter addresses the resulting problem that an RFID reader that receives such an identifier does not know which entity to contact regarding a particular tag and where to get additional information about the identifier and therewith the tag.

The problem is based on the question of whether there exist efficient means to identify oneself to friends while revealing no information to enemies. This is the "Identify Friend or Foe" (IFF) problem. It can be addressed using oblivious transfer protocols [Wei03]. But such protocols are not suited for RFID tags so that a more appropriate proceeding is required.

In the first section, the problem is presented in more detail before the general idea for its solution is presented. Afterwards, pseudonymization infrastructures are introduced and relevant related work in this field is presented. In the subsequent three sections, three approaches, one using asymmetric cryptography and two using hash functions, are presented and discussed. After a section about further optimization by truncating hash values, the chapter concludes with a summary.

5.1 Motivation

In the previous chapter, unstructured tag identifiers were introduced to help addressing the privacy problems in RFID systems: Unstructured identifiers can only be used as unique identifier and no longer contain any additional information. Thus, they do not contain any information about the entity that is responsible for the respective tag and also do not contain any information about the object a tag is affixed to, e.g. information about the manufacturer. To solve the problem with location privacy, methods for varying these identifiers were presented.

But while the unstructured tag identifiers help solving the privacy problems of RFID systems in large part, they have the disadvantage that they limit scalability of the system: This is because structure enables division of responsibility in a simple manner. For instance, tags having *manufacturer #1* in their identifier could be served by a server of *manufacturer #1*, tags having *manufacturer #2* could be served by a server of *manufacturer #2*, and so on. Obviously, one can easily create a well scalable system this way. Using unstructured tag identifiers, such a separation of responsibility is not that simple any more, because there is no information in such an identifier that identifies the responsible entity. This makes the schemes that use unstructured tags only applicable in closed systems of limited size where all the unstructured identifiers can be handled by a single entity.

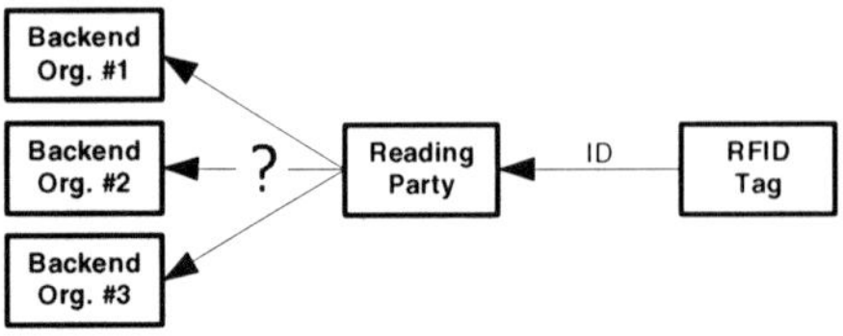

Fig. 5.1. Mapping problem of inter-organizational RFID systems with unstructured identifiers

The goal is to make the schemes that use privacy-friendly unstructured identifiers applicable to open, inter-organizational RFID systems as well. The problem that we face in such systems is depicted in figure 5.1: If a reading party queries a tag, it obtains the unstructured tag identifier. As this identifier does not contain any information about the responsible entity, the reading party does not know who to contact to obtain additional information about the object the tag is affixed to. The only possibility for the reading party would be to contact all known organizations one after another until the organization that knows the identifier and that therewith is the one in charge of the tag is found. Obviously, this is not a scalable solution. A better approach for finding the responsible entity, if an unstructured tag identifier is given, needs to be found.

5.2 Basic Idea for Addressing the Problem

As stated in the previous section, a possibility is required to provide a mapping between a tag and the entity that is responsible for that tag. The constraint is that any information provided by the tags needs to appear completely random for an outsider so that it neither can be abused for tracking nor violates any other privacy goal.

Consequently, the information provided by a tag may not reveal any utilizable information to an outsider but needs to contain all information that is required for a legitimate party to contact the responsible entity.

The idea to solve this conflict is to employ a pseudonymization infrastructure. In this idea, the information provided by a tag is a pseudonym that contains all information to forward a message to the entity that is responsible for that particular tag. The pseudonym must be completely useless for performing any other task than using it as the destination address for message forwarding in the pseudonymization infrastructure. The course of action is as follows: A reading entity contacts a tag and obtains the pseudonym. This reading entity can either be a legitimate party, an illegitimate party, or an attacker with bad intentions. However, the reading entity cannot extract any information from the obtained tag identifier. In the next step, the reading entity creates a request message for querying the entity that is responsible for the tag for further information. This request might contain the identity of the reading entity, the reason for the reading process, the kind of information requested or whatever is necessary for the particular application. The request message is then given to the pseudonymization infrastructure with the pseudonym as destination address. Only the pseudonymization infrastructure is able to extract the information out of the pseudonym that is required to route the message to the destination, i.e. the responsible entity. Upon reception, the responsible entity checks whether the request in the message is legitimate. Based on this check, the responsible entity sends the requested information or a refusal message back to the reading entity. If required, this can be done using an anonymization infrastructure (e.g. [Cha81] or [BFK00]) so that the identity of the responsible party needs not be revealed.

After identification of the requirements and the presentation of related work, different implementation concepts for such a pseudonymization infrastructure will be presented in the following sections. At first, one that is based on asymmetric cryptography which follows closely the concepts used in current anonymization and pseudonymization infrastructures is explained. After that, a pseudonymization infrastructure that is based on one-way hash functions is introduced, improved, and optimized.

5.3 Pseudonymization: Introduction and Related Work

In the previous section the term "pseudonym" was used without further explanation. It means "false name" and is a Greek term. A pseudonym is a representative that is used instead of the real name of an object and is often used to conceal the real identity of a user or a device for privacy reasons in information technology. Besides this application, pseudonyms are often used to ease usage or to increase efficiency. For instance, instead of the real name of a user, mostly a pseudonym, e.g. in this case a shorter "login name", is used for unique identification of that user within a domain. In relational database management systems, often integer values are used as pseudonyms for users or other data to decrease redundancy and increase efficiency.

In communications, pseudonyms are often used as nicknames, e.g. in amateur radio or in instant messaging. Another popular example is lonely heart ads in newspapers in which a "chiffre number" is printed instead of name and telephone number. Whereas in other examples, the intention of using pseudonyms is not primarily to conceal the real name but to provide a more convenient or more efficient means for naming, here chiffre numbers are used to prevent direct linkability. Only a third party, in this case the publisher of the newspaper, is able to link the real name and the chiffre number, i.e. the pseudonym. Obviously, the third party needs to be trusted not to reveal the link between real name and pseudonym to illegitimate parties. As this cannot be presumed in general, one can share trust among several parties. This concept is known as "shared trust" and illustrated in a pseudonymization scenario in figure 5.2. To forward a message to the receiver, a sender sends it to *Node A* with the pseudonym P_A as destination address. The *Node A* maps the pseudonym P_A to another pseudonym P_B and forwards the message to *Node B*. That node maps the pseudonym P_B to the receiver's real name R and can thus forward the message to its ultimate receiver. Only the two infrastructure nodes together have the ability to provide the link between the initial pseudonym P_A and the real name R and thus implement the concept of shared trust.

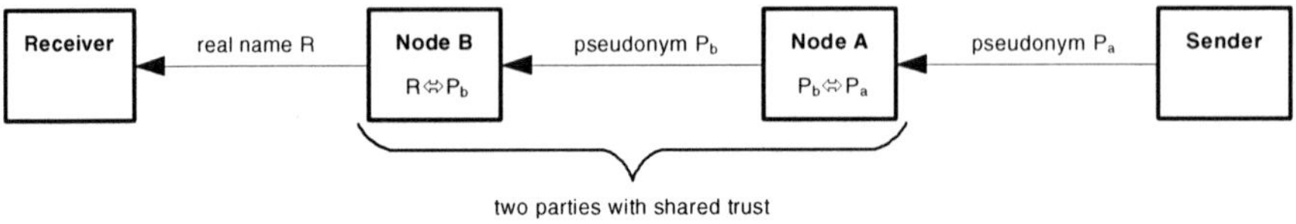

Fig. 5.2. Basic concept of pseudonymization with shared trust

Figure 5.2 does not only demonstrate the concept of shared trust but also depicts a very simple pseudonymization infrastructure in which there is no single party that needs to be trusted. The infrastructure is capable of forwarding messages addressed with the pseudonym P_A as destination to the receiver with the real name R. Some major conceptual disadvantages of that simple pseudonymization infrastructure are that each intermediate node needs to maintain information about each receiver which limits scalability, that the message path is static, and that there is no support for multiple pseudonyms belonging to the same receiver without keeping multiple entries in each intermediate node.

Chaum had the idea to use asymmetric cryptography to implement the concept of shared trust. In 1981, he published a seminal paper [Cha81] in which he described concepts for using pseudonyms for communications over unsecured networks. The core idea presented in that paper set the foundation for today's anonymization and pseudonymization infrastructures.

Chaum's basic intention was to provide a means for anonymization, i.e. he wanted a sender to be able to send messages to a receiver without revealing the sender's identity while concealing the communications relationship to outsiders as well as to the infrastructure itself. Chaum's infrastructure consisted of so called "mixes", i.e. computers that are used for relaying messages. He used these mix servers in conjunction with public-key cryptography to hide the participants and the content of messages and allowed the receiver to send replies to messages without knowing the identity of the sender. The basic idea is to encipher a message with the public key of the receiver, to add the address of the receiver, to encipher all with the public key of an intermediate node, to add the address of that node, to encipher all with the public key of another intermediate node, and so forth (see figure 5.3). This creates a fixed path that needs to be used to deliver a message: Each intermediate node can obtain only the information that is required to discover the next node to send the message to. Thus, only all intermediate nodes together have the necessary information to decipher the routing information and to deliver the message, and only the receiver has the private key that is required for obtaining the message content.

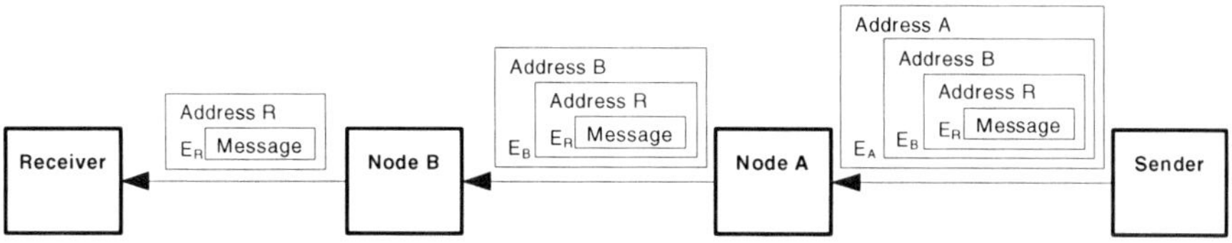

Fig. 5.3. Principle of Chaum's concept

In the 1990th, anonymous remailers that work unidirectional as well as pseudonymous remailers that offer a return path to the sender were built using Chaum's concepts. The "penet remailer" was a service operated between 1993 and 1996 that became quite well-known before it was closed due to legal attacks. Much research was performed to make the systems more secure but also to cope with attacks and abuse, e.g. [MK98].

Goldschlag et al. [GRS99] introduced the term "routing onions" as visualization for the use of multiple layers of encryption. Each intermediate infrastructure node removes a layer of the onion of encryption layers by deciphering with its private key. This way, the routing instructions required to forward the data are revealed whereas the rest of the onion containing data for the other nodes along the path remains untouched.

On the basis of Chaum's work, Goldschlag et al. created "onion routing". This term names the concept of an infrastructure that intends to provide private communications over a public, unsecured network. This means that the infrastructure is not only intended to send single messages but to establish a kind of tunnel for data that supports protocols like HTTP and FTP between communication peers. To achieve this, kinds of "real-time mixes" are used as infrastructure nodes. The intention of the system is that nobody within the infrastructure can tell who is communicating to whom and to keep the sent data private. As it is based on Chaum's concepts, it is also possible for the sender to remain anonymous to the receiver and nevertheless enable the receiver to send replies using so-called "reply onions".

Additional effort was spent to make the infrastructure resistant against traffic analysis: It shall be hidden who is communicating to whom and how often a communication takes place. To achieve this, it must not be possible to trace data that is flowing through the system, e.g. by data size or timing of packet flow. Such issues were addressed in the development of onion routing, for instance by delaying and reordering data in intermediate network nodes, and were subsequently improved by other researchers to guard against advanced attacks, e.g. "intersection attacks" that rely on the fact that nodes can fail or leave the network.

The work of Chaum and Goldschlag became the basis of untraceable remailers (e.g. [DDM03]) and current anonymization infrastructures like Tor [DMS04] or I2P [Web06]. All the systems are built upon the same basic principles: The messages are routed over multiple network nodes, i.e. mixes, with the intention that each node increases the achieved level of protection. The design of the systems ensures that only multiple nodes together can establish a link between the communicating peers. This means that trust is shared among several entities with the intention that even if one or even several of the nodes on the data path share their data with an attacker or if network links are eavesdropped, nobody can link the communicating parties. Besides the security aspects, a main perspective of the current developments is to make the infrastructures more reliable and to provide incentives for users to act as intermediate nodes.

Fig. 5.4. Routing onions as pseudonyms

For being able to send replies without revealing the identity of the sender, Chaum introduced "untraceable return addresses", a concept that Goldschlag made popular with the term "routing onions". These return addresses can be regarded as pseudonyms for the initial message senders. Because of that, the presented infrastructures are not only anonymization infrastructures but pseudonymization infrastructures as well. The same concepts can be used for communications in both directions, so that systems like Tor are able to give both communication peers a pseudonymous identity. An example of the concept is shown in figure 5.4: Different encrypted layers of node addresses on the path can be regarded as pseudonym that stands for a certain destination.

In the following section, the special requirements for pseudonymization infrastructures to be used in connection with RFID systems will be discussed.

5.4 Definition of Requirements and Common Concepts

A pseudonymization infrastructure that shall be used in RFID systems needs to fulfill a number of requirements. Important ones are functionality, reliability, security, resource consumption, and scalability. These shall be discussed in more detail in the following whereas other wanted characteristics like short pseudonyms which lead to a high speed of reading or ease of administration of the infrastructure are only mentioned where appropriate.

FUNCTIONALITY: As already mentioned, pseudonyms may not reveal any information that can be used for tracking, neither directly nor by constellation. Thus, they need to be generated in such a way that they appear to be completely random. Due to the fact that such a pseudonym can act as an identifier itself, it needs to change regularly. This can basically be done in two ways: Either the tag needs to be able to create pseudonyms by itself without interaction with the responsible entity, or the tag needs to be equipped with new pseudonyms by the responsible entity in a secure manner. Both options have advantages and disadvantages compared to each other: The one stated first, i.e. the offline-generation option, is the more straightforward one. It is better scalable since no communication with the responsible entity is required for updating a pseudonym. It can be ensured that the pseudonym changes regularly or even on every read. Another advantage is that no effort is required for securing the communications channel when updating pseudonyms. The other option that involves the responsible entity has the advantage that the tag needs no resources for pseudonym generation. With that option, it is not required to store the identity of the responsible entity in the tag, so that an attacker cannot reveal it even through physical extraction. Within this subchapter, the focus is laid on the offline-generation approach. However, implementing the second approach is also possible with the infrastructures to be presented.

The pseudonymization infrastructure itself needs to have the ability to forward messages with pseudonyms as destination addresses. Therefore, it needs to be able to extract a communication path from the pseudonym whereas an attacker or any other entity shall not have that ability. This conflict can be solved that way that the pseudonymization infrastructure has information (cryptographic keys or other data) that other, potentially hostile entities do not have.

There is functionality that might be interesting for pseudonymization infrastructures in general but that are be considered in the following sections. One feature is concealing the sender so that the receiver is not able to reveal the sender's identity. To achieve this, the sender could provide a pseudonym as source address. Another feature would be to provide end-to-end confidentiality for the messages to be transmitted so that the pseudonymization infrastructure does not get to know the content of the messages it delivers. Besides end-to-end confidentiality, it is necessary to encrypt messages on communication links within the pseudonymization infrastructure so that they appear to be different and to ensure that all the messages routed through the infrastructure have the same size. This is done to prevent that the path of messages can be tracked on their way through the infrastructure by eavesdropping communication links: If the message content or at least the message size remains the same, messages on different links can be linked. Another way to link messages is by using timing attacks which can be counteracted by reordering of messages and by inserting dummy messages. However, such advanced functionality is beyond the scope of the presentation of approaches in the following sections but should be considered in practical application.

RELIABILITY: As the pseudonymization infrastructure is essential for correct functioning of the RFID system, it is vital that the provided service is constantly available. Therefore, the approaches to be presented are designed in such a way that availability of the pseudonymization infrastructures can be ensured without many difficulties. It is considered what measures need to be taken to make the infrastructures reliable, too.

Of course, it is also required that each of the operators of an infrastructure has the intention to provide service. For instance, it is conceivable that an attacker operates a part of the infrastructure. He might deliberately drop messages, create errors or denial service by other means. The only appropriate way to counteract such a behavior is to implement means to detect it so that one is able to sanction it. This could be done by implementing a mutual control of different operators. However, such considerations are beyond the scope of this text.

SECURITY: Security is one of the main design criteria in a pseudonymization infrastructure because the main intention of the latter is to ensure that a pseudonym does not allow conclusion to its owner. Because of that, security issues will be the main focus of discussion of the approaches to be presented.

All the presented pseudonymization infrastructures are based on the concept of shared trust. They consist of different infrastructure nodes and only multiple nodes together can perform the delivery of messages. This has the advantage that there is no central authority that needs to be trusted. This is therewith a completely different approach to the pseudonym protocol presented by Molnar et al. in [MSW05] that uses a central "Trusted Center".

Ideally, the infrastructure shall appear as a black box to sender, receiver, and arbitrary outsiders. It shall not reveal any exploitable information to an outside observer. An attacker can have different capabilities. Depending on the weaknesses of a certain approach these capabilities might present a security problem or not. Because of that, different attacker capabilities and different targets for attacks are presented in more detail in the next section.

RESOURCES: It is a design goal of the approaches to be presented that as few resources as possible are utilized. Resource consumption is especially relevant for operations that need to be performed by tags as well as for the amount of data that needs to be stored on tags. Because of that, expensive cryptographic operations should be avoided. Pseudonyms should be short, i.e. have a size below 1 Kbit, so that they do not need too much memory and do not occupy the communications link between reader and tag for too long a time.

Within the pseudonymization infrastructure, resource consumption is not such an important issue as in tags. It is nevertheless desirable to save resources to keep operating costs for the pseudonymization infrastructure low and make it scale more easily.

SCALABILITY: It is a challenge to create a scalable pseudonymization infrastructure, i.e. one that is able to process millions of messages per second. One measure to achieve this is to use a distributed system so that load can be spread. Because of that, it is not only a security measure to do without a central authority but also one regarding scalability.

For scalability reasons, it is also advantageous to design the message routing in such a way that the nodes in the pseudonymization infrastructure do not need to maintain state information for individual messages, for instance for error handling purposes.

In a pseudonymization infrastructure that shall handle a high number of possible receivers, it is essential that adding new receivers is a simple task that does not involve a high number of nodes of the distributed infrastructure. Because of that, the approaches to be discussed are designed in such a way that not more than a single (logical) node needs to be updated when a new receiver shall be served by the pseudonymization infrastructure.

To achieve better scalability and to make mirroring of nodes (for load sharing or increasing availability) easier, writing database access in nodes should not be needed except for adding a new receiver or adding new nodes to the infrastructure. All of the pseudonymization infrastructures in this chapter are designed with this in mind.

5.5 Attack Targets and Attacker Capabilities

The pseudonymization infrastructure to be created needs to fulfill three main purposes: It shall hide the identity of the entity that is responsible for a tag. On the other hand, it shall provide a link between a tag and the responsible entity so that the reading entity has a destination for issuing a request for further information regarding the tag. The third purpose is to prevent tracking by the pseudonyms issued by a tag.

ATTACK TARGETS: Any means to counteract the purposes for which the pseudonymization infrastructure is designed is an attack target. The following attack targets are examined:

- *Gathering the identity of an entity belonging to a given pseudonym*
  An attacker has the intention to get to know which entity a pseudonym belongs to. In the RFID scenario, this means that an attacker gets to know who is in charge of a tag.
- *Confirming the identity of an entity belonging to a given pseudonym if the attacker has a guess of that identity*
  Compared to the previous target, an attacker has a guess to which entity a

pseudonym belongs. Then, his task is to confirm that his guess is right. Generally, this is easier to achieve than gathering the identity of the entity to which a pseudonym belongs.

- *Tracking or recognition of a tag despite of changing pseudonyms*
  The efforts for ensuring security and privacy that are taken for RFID systems in this book intend to provide location privacy and to prevent unwanted profiling. This can only be achieved when all parts of the system are properly designed. It is a target for an attacker to bypass these safeguards. This can be done if he is able to prove or guess that different pseudonyms belong to the same entity.

The protection against tracking is an optional feature in a pseudonymization infrastructure because such an infrastructure can be used in two ways: The first and simpler way, the receiver generates a pseudonym and gives it to the sender instead of his real address. The intention is here to hide the receiver's real identity from the sender and the network. As the generation of the pseudonym is done by the receiver, no burden is put on the sender. This makes this option easily implementable in the RFID scenario because a tag would not require computational resources for pseudonym generation. But on the other hand, location privacy is not assured because the pseudonym could be used for tracking purposes.

The second way that was already motivated in the section 5.4, the sender creates pseudonyms by itself and uses different pseudonyms for sending messages to the receiver. For this, the sender needs to have enough information about the receiver to be able to do this. The intention of this approach is to conceal the communications relationship to attackers that are eavesdropping or operating one or more hops of the communications path. The sender has the burden to generate pseudonyms and needs the computational resources for doing this, but on the other hand, the changing pseudonyms cannot be abused for tracking, which makes this approach a good match if location privacy in an RFID scenario is an issue.

In the following sections, the second, more complex option including the protection of location privacy will be considered. However, the schemes are also applicable in the other case. Then the measures for preventing tracking are not required.

ATTACKER CAPABILITIES: Depending on the design of the pseudonymization infrastructure, an attacker needs certain abilities to be able to perform a successful attack. It makes sense to distinguish the following capabilities for the considerations in the next sections:

- *Eavesdropping a single pseudonym on the link between sender and root node*
  This is the most probable capability. In the RFID scenario, it corresponds to an attacker's capability to operate a reader by himself or to eavesdrop the wireless communication between a tag and a reader. It can be assumed that an attacker always has this capability if he has one or several of the other ones listed here.
- *Support within pseudonymization infrastructure*
  An attacker can have support within the pseudonymization infrastructure, e.g. one or several infrastructure nodes that provide information to him.

- *Data collection*
  This capability means that an attacker can obtain information regarding more than a single pseudonym or message of a fixed sender. Obviously, in this case, he has more raw data that can be used for exploits compared to the case that he has only information about a single pseudonym or message. One could distinguish here whether the attacker or a group of attackers alone perform data collection or whether there is help from intermediate nodes or receivers, but this is out of scope of the considerations here.
- *Physical extraction*
  This capability denotes that an attacker can access information that is available to the sender but that is not public. In the RFID scenario, an attacker could access a tag's memory using physical extraction (see sections 3.2 and 3.5) and access private data that is used in the pseudonym generation process.

5.6 Approach Based on Asymmetric Encryption

In this section, an infrastructure for pseudonymous messaging that is based on asymmetric encryption is presented. It closely follows the concepts known from onion routing and thus inherits many of its properties. The idea is to use routing onions (see figure 5.4) whose layers contain names of intermediate nodes and random numbers as pseudonyms.

Topology and setup

The infrastructure consists of an arbitrary number of intermediate nodes and a designated node that is called "root node" in the following (see figure 5.5). The intermediate nodes work as mixes. They are independent and do not have a special structure among each other. The root node provides the entrance to the infrastructure, i.e. it is the node that is contacted when a message shall be forwarded through the infrastructure. For this, a single node is required because otherwise the address or another means for identifying the first node to be contacted would have to be included in a pseudonym. But this may not be done since such a selector could be abused as a means for tracking by constellation.

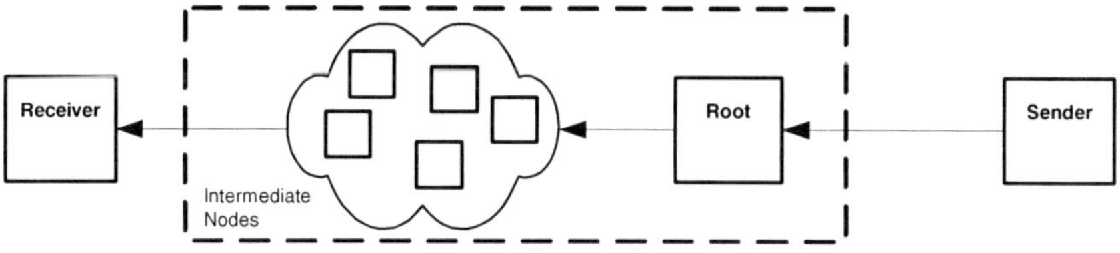

Fig. 5.5. Infrastructure topology for approach based on asymmetric encryption

All intermediate nodes as well as all receiver nodes each have a unique identifier that provides the information how to reach the node in the network. This could in its simplest form be the DNS hostname of the node. All the nodes generate a key pair of an asymmetric encryption scheme. The public keys are published; the corresponding private keys are kept secret.

Path provisioning and pseudonym generation

The receiver selects a number of intermediate nodes, e.g. two in figure 5.6, as mixes. The identifiers of these nodes, the identifier of the receiver node, and all the corresponding public keys (one from the receiver and the ones of the intermediate nodes and the root node) are then given to the sender, i.e. in the RFID scenario the public keys are stored in the RFID tag. With this information, the sender is able to generate pseudonyms as follows: The sender takes a random number r and the identifier of the receiver node and enciphers both using the public key of the last node before the receiver. In the next step, the sender enciphers the ciphertext of the previous step and the identifier of the next intermediate node. This step is repeated for each intermediate node. The final ciphertext is the pseudonym and contains all the information that is required for delivering a message with the pseudonym used as the receiver's address.

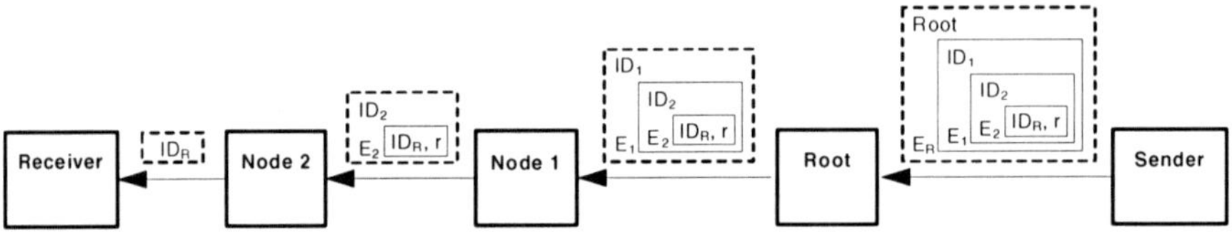

Fig. 5.6. Pseudonyms including random numbers

The random number creates diffusion and makes the pseudonyms look different even when designating the same receiver. It works as a so-called "salt". $\#P = 2^{d_r}$ different pseudonyms can be created for the same receiver. In this formula, d_r denotes the domain, e.g. $d_r = 16$ means that the random number has a 16-bit domain. Another way to create different pseudonyms for the same receiver would be to give a higher number of intermediate nodes than the number of nodes to be used to the sender. The required number could be selected out of them and so different paths through the infrastructure and thus different pseudonyms be created. For example, if $n = 6$ intermediate nodes were known to the sender and the path should contain $(t - 1) = 2$ intermediate nodes (t is the depth of the node tree that forms the infrastructure) as in figure 5.6, then $p = n!/(n - t + 1)!$, i.e. here in the example $p = 6!/4! = 6 \cdot 5 = 30$, different paths and thus pseudonyms could be created. Obviously, either a high number of intermediate nodes or a high depth of the infrastructure is required to gain such a high number of possible pseudonyms like when using the scheme with the random numbers. But storing the identifiers and public keys of a high number of possible intermediate nodes in the RFID tag would require much memory so that this approach is not appropriate and the previously described approach of creating diffusion by inserting random numbers is preferred.

Message forwarding

If the sender wants to forward a message to the receiver, he attaches the generated pseudonym and forwards the message to the root node (see figure 5.6). The latter removes the outer layer of encryption of the pseudonym by deciphering it using its private key. Then the part of the pseudonym which is encrypted with E_2 is used as new pseudonym. Then the message is forwarded to the destination that is denoted by the decrypted identifier ID_1. At the intermediate nodes, the steps described for the root node are performed analogously. The last intermediate node removes the last layer of encryption, discards the random number r, and forwards the message to the ultimate receiver.

Enhancements

Using hostnames as node identifiers makes the pseudonyms become quite long: The hostnames of the intermediate nodes and the receiver need to be stored in the pseudonyms. Since the pseudonyms need to be of fixed length to prevent a means for tracking by constellation, even if the hostnames are short for special nodes, space needs to be reserved for the longest hostname that is allowed in the system. This would mean that if two intermediate nodes were used (and thus three hostnames would need to be stored) and the maximum length of allowed hostnames was defined to be 64 characters, the pseudonyms would have a size of $3 \cdot 64$ byte = 192 byte plus the space needed for the random number plus, if required, padding for filling the blocks of a block cipher.

Because of that, it makes sense not to use hostnames as identifiers but numbers that act as an index in an extern, publicly available node table. This measure extremely decreases the size of pseudonyms: If 32 bit numbers were used as index and a 32 bit random number were used for creating diffusion, the layer with the random number would have a size of 64 bits and all other layers a size of 32 bits. A ciphering scheme that is based on blocks of 64 bits and a path with the root node plus two intermediate nodes leads to pseudonyms with a size of $3 \cdot 64$ bit = 192 bit = 24 byte which is only one eighth of the pseudonym size shown in the example above without this optimization.

Besides enciphering only node identifiers and random numbers, it would be possible to encipher the actual message as well and thus to guarantee its confidentiality. This feature is not considered further because it is an optional one and not part of the stated requirements to be fulfilled.

Discussion

The scheme presented in this section closely follows the concept of routing onions. Due to the layered encryption which can only be undone by cooperation of all chosen intermediate nodes, the previously selected path through the messaging infrastructure needs to be followed to deliver a message. Each of the nodes on the path can only derive from the current pseudonym to which node the message needs to be sent next so that the scheme implements the concept of shared trust. The scheme is simple and comprehensible.

RELIABILITY: The selection of intermediate nodes to be used and making this information available to the sender takes place before pseudonyms are generated. Thus, it is required that all these nodes on the message path are available for transmitting a message to the receiver. A rerouting of the message path over other nodes is not possible since the required information to do this is not present in the nodes for security reasons.

To achieve availability, several measures are conceivable. To minimize downtime of a node, it should have a redundant network connection and should be hosted on at least two machines, so that the provided service remains available in case one machine fails. For transient failures like link failures that require some time for rerouting in the network, it also makes sense to queue messages in a node in case that the next node on the path is currently unreachable and retry to transmit the message some time later. To avoid congestion of a network segment by repeated retries, an exponential back-off like the one done in TCP [Ste94] should be performed in this case.

Besides these measures, one can equip the sender with a larger set of intermediate nodes than just a minimal number of them. This way, the sender can select a defined number of nodes out of this set so that different paths through the infrastructure can be composed instead of having just a single, preselected path. The disadvantage of this measure is that more memory is required in the sender for storing the alternatives. Note that there can be no alternative for the root node; at least one mirror of it needs to be reliably available.

If a message cannot be transmitted for whatever reason over a chosen path, there are two possible options: The first option is to silently discard the message. This is simple but makes it difficult to track down errors. The second option is to return error messages to the sender. In the latter case, one should consider whether it is preferable to directly send the error back to the sender or to record the traversal of intermediate nodes in the message, i.e. creating a "reply onion", and thus to be able to return the error message on the same path. Sending errors back on the direct way reveals information about the infrastructure, in the other case, the infrastructure appears as "black box" to the sender because he only sends and receives messages via the root node.

SECURITY: A pseudonym that has been created in the stated manner does not reveal any information about the receiver to an attacker: All the private keys of the intermediate nodes are required to decrypt the information about the receiver that is stored in the pseudonym. Thus, the protection of the receiver's identity lies in the algorithm being used for encryption. If only short identifiers and random numbers are encrypted, there is even not much redundancy that could help in cryptanalysis.

Pseudonyms are independent from each other: If an attacker is in possession of one or many pseudonyms, he is not able to create another pseudonym unless he is capable to break at least the outer layer of encryption. Having more than a single pseudonym, for instance by eavesdropping a sender's communication for a longer time, does not make it easier for an attacker to reveal the identity of the receiver.

Hence, the pseudonyms created in this approach based on asymmetric encryption are very secure as long as the private keys remain secret. The most important key in the system is the private key of the root node. As the outer layer of all the pseudonyms is encrypted with the public key of the root node, knowing the corresponding private key enables to decipher the outer layer of all pseudonyms and thus to get to know the first intermediate node. If the pseudonyms were created in such a way that always the same path through the infrastructure to a particular receiver is used, an attacker would become able to perform tracking by constellation: Then a fixed set of senders always returns pseudonyms with the same set of first intermediate nodes. Because of that, more confidence is put on the operator of the root node than on other ones.

RESOURCES: The sender needs a lot of computational resources to perform the asymmetric encryption operations that are required for creating pseudonyms. Even if tags had the capability to perform such operations, too much time would be required for processing to achieve the speed of reading of tags that is demanded by industry. Consequently, the scheme is not suited for application in tags – neither low-cost ones nor more expensive ones.

The asymmetric encryption operations also put a high burden on intermediate nodes. This limits the number of messages that an intermediate node can process within a particular period of time considerably. Due to today's fast CPUs and the possibility to distribute load among several machines, this issue can be easily addressed.

SCALABILITY: If there is a means to update the node information of many senders, the number of intermediate nodes can easily be increased so that the load of a high number of messages can be spread onto many machines. Since all messages are routed through the root node, this node is the bottleneck of the infrastructure. To get along with a high number of messages, it is required to mirror the root node so that the load is distributed onto more than a single machine. Doing this is simple because each machine that is in possession of the root's private key can act as root node. No further mirroring is required because changes in the infrastructure and adding of new receivers are completely transparent to the root node.

Distributing the load among a number of intermediate nodes or by mirroring single nodes is not a problem in the scheme since no cooperation or coordination between nodes is required: Adding or removing intermediate nodes or adding or removing receivers is transparent to other infrastructure nodes so that no additional resources are required for synchronization tasks. This makes the approach flexible and well scalable.

5.7 Basic Approach Based on Hash Functions

In the previous chapter, one-way hash functions have proved to be an adequate primitive to implement safeguards against counterfeiting of RFID tags and unwanted recognition and tracking. Starting from this, it is self-evident to try to implement the needed pseudonymization infrastructure by relying on the same primitive.

Obviously, a one-way hash function is less "powerful" than asymmetric cryptography because the latter is an operation that can be performed in both directions in case the appropriate keys are available. Because of that, the prospects of being able to implement an infrastructure that yields the same features and level of security as a one based on asymmetric cryptography are quite poor. Nevertheless, as will be shown in the following, implementing an infrastructure based on one-way hash functions that is adequate for the use in the RFID scenario is possible in principle.

It makes sense to adhere to the concept of shared trust and thus to use multiple intermediate nodes like in onion routing. Similarly to onion routing, the concept needs to ensure that each node can only derive that information from pseudonym data that is required for getting the address of the next node along the path. Security must obviously be based on the one-way characteristic of the used hash function. The main idea is to differentiate a legitimate node and an attacker by the amount of information that is present: Using the asymmetric encryption approach, the legitimate node is in possession of the needed private key for deciphering whereas an attacker is not. This can be transferred to the one-way hash function approach: A legitimate node is in possession of data in a database that is used to reduce search space or to perform calculations whereas an attacker is not and thus needs to search the entire search space to derive some information from a hash value.

Within a project thesis [Wei05], a simple pseudonymous remailer was implemented by a student to demonstrate that the concept presented in this section can be used in practice. This illustrates the feasibility of the approach. The idea and the basics of using hash functions as a primitive for the creation of pseudonymization infrastructures has also been published in [HGM06a] and [HGM06b].

Topology and setup

Like in the asymmetric encryption approach, the entry point into the infrastructure needs to be a single node which is called "root node" (see figure 5.7). Otherwise, an identifier would be required as a node selector. But such a selector would be a means for identification itself and thus could be abused by an attacker for illegitimate tracking.

The root node has a number of child nodes, each of which has a number of child nodes itself, and so forth on the next layer. Receivers are the child nodes of the lowest layer nodes of the pseudonymization infrastructure. This way, a hierarchical tree topology is created which is similar to the tree topology found in the domain name system (DNS, see [Moc87]) which has proved to be powerful and well scalable. The tree topology has the advantage that the number of child nodes of each node can be

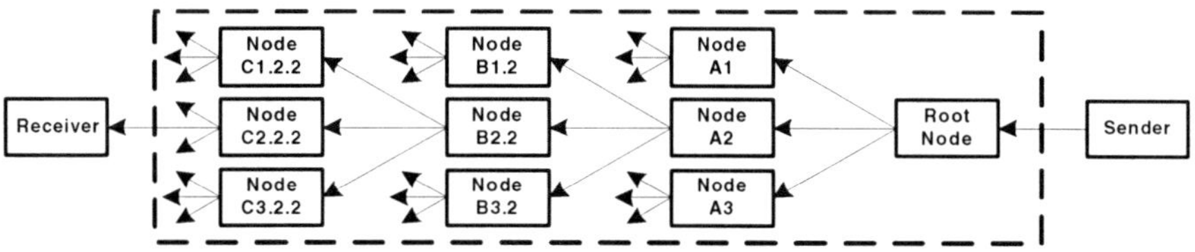

Fig. 5.7. Topology of the basic hash-based infrastructure

in similar order of magnitude and that the resulting structure is thus balanced and homogeneous.

Each of the nodes in the infrastructure assigns a unique identifier to its child nodes. The identifiers are denoted by N in the following and should be chosen in such a way that they cannot be guessed. A possible algorithm for creating identifiers is to concatenate the respective node's hostname with a random number and finally to calculate the hash value of that string using a one-way hash function. The mapping between N and the corresponding node's address is not revealed to any other node or even to the public. Each node has a database table that links the node identifiers N_i of its child nodes to the hostnames of the child nodes.

Each node in the pseudonymization infrastructure also has a database table in which record rows in the following form are present:

$$h(k, N_{\text{child}}) \rightarrow k, N_{\text{child}} \quad \forall N_{\text{child}}, k \in [0, r_{\max}[\subset \mathbb{N}_0$$

h denotes the one-way hash function. The parameter $r_{\max}$ is a natural number that is used to create a trade-off between the number of different pseudonyms that can be created for a particular receiver and the space that is required in the database. As $r_{\max}$ records are stored for each child node, $\#N_{\text{child}} \cdot r_{\max}$ rows need to be stored in the database table. A higher $r_{\max}$ increases the number of possible pseudonyms and therewith the level of privacy the infrastructure is able to offer. If $r_{\max}$ is identical for all nodes in the path, then $n = r_{\max}{}^{t}$ is the number of different pseudonyms that can be created for a particular receiver whereby t is the depth of the tree topology, i.e. the number of nodes on the path (intermediate nodes plus receiver) less the root node, so that $t = 4$ in figure 5.7. As n grows exponentially, t should be five or more in practice. $r_{\max}$ should be at least 10^5 to get a reasonably high result without putting to much burden on individual nodes. For the root node and on the first level, it should preferably be an order of magnitude higher to get a better protection against tracking by constellation.

Path provisioning and pseudonym generation

When a receiver joins the infrastructure and therewith becomes a leaf node of the tree topology, the path through the pseudonymization infrastructure from the root node to the receiver node becomes set and does not change any more. Besides the root node, the path consists of a number of intermediate nodes. On joining the infrastructure, the receiver gets the node identifiers N_i that describe this path. Note that the receiver does not need the information to which physical nodes these node identifiers correspond to.

The node identifiers N_i are the information that is required to create pseudonyms. The receiver can use them to create pseudonyms on its own or can give these identifiers to any entities that shall be able to create pseudonyms. These entities should be trusted since the security of the system decreases considerably with knowing the node identifiers. In the RFID scenario, the RFID tags shall create pseudonyms and thus need to obtain the node identifiers N_i corresponding to the path to the respective receiver for doing so.

For creating a pseudonym in an infrastructure of depth t, the node identifiers $N_t..N_1$ as well as t random numbers $r_{t-1}..r_0$ are required. Then a pseudonym P is a vector with t elements:

$$P(p_{t-1}, p_{t-2}, \ldots, p_0)$$

The elements are calculated using the following rule:

$$p_i = f_{r_0}(f_{r_1}(\ldots f_{r_{i-1}}(h(r_i, N_{i+1}))\ldots))$$

The random numbers r_i need to be between zero (inclusive) and $r_{\max}$, i.e. $r_i \in [0, r_{\max}[\subset \mathbb{N}_0$. p_0 degrades to $p_0 = h(r_0, N_1)$, i.e. the hash value of the random number and the node identifier of level below the root node. h again is the one-way hash function. f_r is an invertible function that scrambles its preimage using the parameter r. It can be based on a simple bitwise xor operation: $f_r(x) = x \oplus s(r)$. s is calculated using the formula:

$$s(r) = r \cdot {r_{\max}}^j + r \cdot {r_{\max}}^{j-1} + \ldots + r \cdot {r_{\max}}^1 + r = r \sum_{i=0}^{j} {r_{\max}}^i$$

The parameter j is chosen in such a way that the domain of s is greater than or equal to the domain of x but as small as possible within that restriction. If $r_{\max}$ is a power of 2, i.e. $r_{\max} = 2^b$, then calculating f_r can be implemented very efficiently using bitwise xor operations and binary shift operations only:

$$s(r) = r \sum_{i=0}^{j} {r_{\max}}^i = r \sum_{i=0}^{j} (2^b)^i = r \sum_{i=0}^{j} 2^{(b \cdot i)} = \bigoplus_{i=0}^{j} [r \ll (b \cdot i)]$$

This means that $s(r)$ becomes the bitwise concatenation of several occurrences of the random number r each of which is represented with b bits. The operator $\ll$ in the formula denotes a bitwise left shift.

Each element of the pseudonym vector denotes a destination to which a message needs to be forwarded next. p_0 is processed by the root node and denotes the node on the hierarchy level below the root node; p_{t-1} is processed by the parent node of the receiver and denotes the node of the receiver. To be able to interpret an element p_i, information from all the nodes on higher levels of hierarchy is required:

The random numbers r_i of all the higher levels are required to invert the repeated applications of the function f. This way, the concept of shared trust is implemented. The nearer a node on the path to the receiver is, the more is the corresponding hash value of the node identifier in the element of the pseudonym vector scrambled by repeated applications of the function f with random numbers that are not known to any intermediate node of the infrastructure or to the receiver.

The random numbers have thus two objectives: On the one hand, they make it possible to create different pseudonyms for a single receiver. On the other hand, they make the nodes depend on each other to derive the hash values.

Due to the hash calculations, each element p_i of the pseudonym vector has the length of the output of the employed hash function, e.g. 128 bit for MD5 or 160 bit for SHA-1. The function f does not change the length because it provides an invertible one-to-one mapping so that domain and co-domain have the same size.

Message forwarding

For sending a message to the receiver, the sender forwards it to the root node with a pseudonym to indicate the final destination. An attacker that overhears this communication cannot obtain exploitable information from the pseudonym because additional background information is required for doing so. The simplest element of the pseudonym vector is the element p_0, which was calculated by the sender using the formula $p_0 = h(r_0, N_1)$. As the hash function is a one-way one, r_0 and N_1 cannot be obtained if only p_0 is given. The situation becomes even more obfuscated for the other elements that are not only calculated using the one-way hash function but also scrambled using one or several random numbers.

The root node now uses its database table that contains all the possible combinations of random numbers and node identifiers to obtain the r_0 and N_1 belonging to the given p_0. An attacker cannot build such a database table on its own as long as he does not know the node identifiers that exist on the current level of hierarchy. In the next step, the root node can look up in its database table of child nodes to which node the node identifier N_1 belongs and therewith knows to which node it needs to forward the message next. Before forwarding the message, the root node inverts the outer f_{r_0} functions of the other elements in the pseudonym vector. For doing so it applies the inverse function $f_{r_0}^{-1}$. The element p_0 is dropped off the pseudonym vector because it is no longer required. This results in a pseudonym vector that consists of one element less than before.

The procedure is performed analogously at the other nodes within the path, each of which treating the respective first element of the pseudonym vector. The last node within the pseudonymization infrastructure inverts the hash value of the last element in the pseudonym vector using its database entries and thus obtains to which receiver the message needs to be forwarded ultimately.

Enhancements

Within the pseudonymization infrastructure, message forwarding takes place between a node and its respective child nodes. It makes sense to encrypt these communication channels, for instance using SSL/TLS [DA99], so that an attacker cannot obtain information regarding the content of the message transfers, especially the pseudonym vectors. This is a good measure to minimize the target area for attacks.

The transmission of the complete pseudonym vector on the wireless interface of the RFID tag needs not be secured. The system is designed in such a way that attackers can get this information without having any possibility to abuse it. This needs to be the case because the attacker cannot only be an eavesdropper of a tag reading but can also be an operator of a reader so that he can obtain the pseudonym vector albeit the wireless transmission being secured in the communication process.

As stated in section 5.4, providing confidentiality for the actual message is not a goal for the infrastructure. Nevertheless, securing the message content is possible in principle. One option is to encipher the links between infrastructure nodes as stated above. This provides good protection against eavesdroppers on communication links, but the messages are still in clear within the intermediate nodes. A second option that solves this problem would be to encipher the message by the sender with the public key of the receiver. Obviously, the sender needs to be able to perform asymmetric encryption for doing so. This would require additional effort and lessen the advantages of the hash-based scheme. Unfortunately, the asymmetric encryption operation cannot be sourced out to the reader in the RFID scenario because the required public key for doing so would be a means for identifying the receiver and could thus be abused for tracking purposes. Another problem would be that the air interface between tag and reader needed to be secured.

Discussion

The presented approach based on one-way hash functions was implemented in such a way that it closely follows the principles of the approach based on asymmetric cryptography: The concept of shared trust is taken over. Database tables with hash values act as a secret analogous to a private key. Due to the limited capabilities of hash functions compared to invertible cryptography, it was proximate to use a more structured tree topology instead of a pool of nodes behind the mandatory root node.

RELIABILITY: Reliability of such a hash-based pseudonymization infrastructure is similar to that of an infrastructure that is based on asymmetric cryptography (see the previous section): The nodes along the given message path need to be available for a message transmission to succeed. Analogously to the asymmetric cryptography case, a rerouting over other nodes is not possible. Measures to achieve reliability are to increase availability of nodes, for instance by redundant network connections, and to mirror nodes for availability and also load-sharing reasons.

For the pseudonymization infrastructure that is based on asymmetric cryptography, it was proposed to provide the sender with a higher number of intermediate nodes so that different paths can be composed. This is not possible in the described hash-based case since the message path through the infrastructure is set when the respective receiver joins the system. Nevertheless, it would be possible for a receiver to join the system multiple times. When the lowest layer nodes are selected in such a way that different nodes are on the message path in the tree hierarchy, the sender can be equipped with an alternative path for backup. Then only the root node needs to be always available since it is always the first node to be contacted by a sender.

SECURITY: Within the presented approach, a pseudonym is a vector that is composed of one element for each node to be traversed on message delivery. Each element is a hash value and except for the first one additionally scrambled using one or more parameterized invertible functions. Alone with one or many of such pseudonyms, an attacker is not able to reveal information about the respective receiver. Without the information that is stored in the database tables of the infrastructure nodes, there is no possibility for abuse.

The random numbers enable the creation of different pseudonyms for the same receiver so that unwanted tracking using a pseudonym is not possible. If random numbers in the range of $0 \le r_i < r_{\max}$ are used on each level of hierarchy, there exist $r_{\max}$ different elements in the pseudonym vector that denote the node on the level below the root node, ${r_{\max}}^2$ different elements that denote the node on the next lower layer and so on. For the last element in the pseudonym vector that denotes the receiver's node, ${r_{\max}}^t$ possibilities exist. If $r_{\max}$ is high enough, even tracking by constellation is successfully prevented.

As long as no further information is available, it is not possible to link different pseudonyms denoting the same receiver without knowing the random numbers used in the generation process and inverting the applied hash functions. This results in the pseudonyms being independent from each other. If one or many pseudonyms denoting the same receiver are given, it is not possible for an attacker to create other pseudonyms.

Due to the stated reasons, the pseudonyms themselves reach a similar security level as the pseudonyms in the approach using asymmetric cryptography. None of the stated attack targets can be reached as long as no further information than a set of read or eavesdropped pseudonyms is available. Thus, the presented scheme provides a good protection of the insecure wireless communication between RFID tags and readers as well as against malicious reading entities.

The major problem of the scheme lies in its static topology and the static node identifiers. It cannot be expected that these identifiers can be kept secret for a long time: All the entities that shall be able to create pseudonyms need to know the node identifiers belonging to the child nodes along the path through the pseudonymization infrastructure. As the node identifiers play a crucial role, they may not have a very limited domain and be created in such a way that they cannot be guessed.

A viable attack is the following: An attacker can join the pseudonymization infrastructure many times as a receiver and therewith obtain the node identifiers of the corresponding paths. By doing so, the attacker has a good chance to get many of the child node identifiers of the root node due to their limited number. With these identifiers, the attacker can create database tables that are identical to the ones in the root node and is thus able to find out to which child node a pseudonym would be routed by the root node. This enables an attacker to perform tracking by constellation. The more node identifiers an attacker obtains, also of the child nodes on the next lower level, the more specific and thus better his tracking capabilities become. The problem increases if many attackers share their findings.

Another attack is to obtain the node identifiers of the nodes along the path to a specific receiver by means of penetrating the sender. In the RFID scenario, this means that the node identifiers of a path to a receiver can be obtained by physical extraction out of a tag. With these identifiers, it is possible to build the database entries belonging to all of these node identifiers. With these entries, it is possible to determine whether a given pseudonym (read or eavesdropped) denotes the same path and therewith belongs to the same receiver or not. This information can obviously be abused for unwanted tracking. Here, the hash-based approach is weaker than an approach using asymmetric cryptography because having the node identifiers, i.e. the data required for creating pseudonyms for a particular receiver, would not enable linking of different pseudonyms.

But note that an attacker having the node identifiers has only the "logical" identity of the receiver. Without knowledge to which physical nodes the node identifiers belong, the attacker cannot determine the network address and therewith the "physical" identity of the receiver. However, due to the static tree structure of the pseudonymization infrastructure it may not be assumed that these mappings cannot be revealed with the help of nodes within the infrastructure or by data collections that are obtained by purposeful eavesdropping of communication links within the infrastructure.

RESOURCES: The presented approach requires one hash calculation for each node on the path to the receiver to create a pseudonym. Assuming that performing a hash calculation requires fewer resources than asymmetric encryption, this is a significant saving of resources. But the scheme demands still too many calculations to be done by low-cost tags in a reasonable time so that its application is limited to more expensive tags.

For intermediate nodes, the presented scheme is much more efficient than the approach based on asymmetric cryptography: Instead of requiring to perform a decryption operation using the private key of a node to forward a message, the node simply has to perform inversion of the function f and to perform two database lookups, one in the table with the hash values and one in the table in which the child node identifiers N_i are mapped to network addresses. The possibility to distribute load among several machines is possible in both infrastructure approaches.

For each child node of a node, there need to be $r_{\max}$ entries in the database table that is used to invert the hash functions. The higher the parameter $r_{\max}$ is chosen, the more space is required in the database. A higher number of $r_{\max}$ provides a better protection of privacy because then a higher number of different pseudonym vectors that denote the same receiver exists. Here, an appropriate balance needs to be selected.

SCALABILITY: Due to the static hierarchy within the pseudonymization infrastructure with fixed paths to a respective receiver, it is not a simple task to add additional intermediate nodes so that increasing message load of a fixed number of receivers can be better distributed. But it is possible to add additional branches to the tree structure so that the number of receivers that can be handled can be easily increased. In the approach using asymmetric cryptography, one is more flexible because senders can use different paths to reach a receiver and can thus better distribute load.

Besides these differences, the hash-based approach has the same scalability issues and features as the approach based on asymmetric cryptography: All messages are routed through the root node so that it is a bottleneck, and the load needs to be distributed onto several machines. Distributing load onto several machines can be easily done: The machines that shall act as a mirror to a node only need the node identifiers of the respective node's child nodes and the corresponding network addresses so that the database tables that are required for forwarding the messages can be created. Adding and removing of receivers does only affect the leaf node of which the receiver becomes (or is, respectively) a child. Write operations on the database tables are not required in any other intermediate node. Mirroring of nodes can thus be done efficiently. All this makes the approach well scalable.

5.8 Advanced Approach Based on Hash Functions

From the security perspective, the approach presented in the previous section has some deficiencies that mainly result from the static nature of the infrastructure: A particular path corresponds to a particular set of node identifiers. The latter are the only secrets in the system and only one for each node exists. Since they are required for pseudonym generation and due to the hierarchical topology only few exist near the root node, the security of the infrastructure cannot be assured in the long term. In this section, the identified deficiencies shall be systematically counteracted while retaining the positive properties.

Main ideas

The approach presented in the previous section uses a unique node identifier for each child node. Variability is ensured using a wide range for random numbers. One idea is to use several different node identifiers for each child node so that the number of node identifiers in the infrastructure is considerably increased. To keep the number of records in the databases in the same order of magnitude, one can limit the range of random numbers accordingly, e.g. to $r_{\max} = 2^8$.

One can further optimize the proportion between node identifiers and random number range depending on the position of a node along the path between sender and receiver: The nodes near the root node have the principal purpose of preventing tracking by constellation so that variability introduced by the random numbers is important there. But near the receiver, additional variability is not needed. The main purpose of the nodes there is to conceal the relationship between pseudonyms and corresponding receiver.

A huge problem within the approach presented in the previous section is that the database tables of the nodes can be built to a large extent when the node identifiers of the children of the node in question are known. There are no additional secret elements because they would need to be given away for pseudonym generation. Based on the idea of an increased number of node identifiers, one no longer needs to stick to a tree topology and each receiver does no longer need to be represented by a unique set of node identifiers representing the corresponding message path. This way, several paths to each receiver can exist, and each path can be composed of different node aliases. This enables the introduction of additional private elements: To each node identifier alias an additional randomly generated private value gets assigned. As a node can have different parent nodes now, each node can maintain a list of valid parent nodes and get assigned a private value for each parent node. So, if two nodes communicate, a private value for the used node identifier alias of the child node exists as well as a private value in the child node that is dependent on the parent node. Both private values together form an additional secret that is dependent on the path and the used node identifier alias. As different combinations form different secrets, this makes it much more difficult to attack the infrastructure.

Topology and setup

For the advanced hash-based approach, the intermediate nodes no longer form a tree topology. Instead, there is more flexibility: An intermediate node may have several parent nodes. Apart from that, the advanced hash-based pseudonymization infrastructure is formed similarly to the basic hash-based approach. In figure 5.8, a topology overview in which, for the sake of clarity, each intermediate node has a single parent node is depicted.

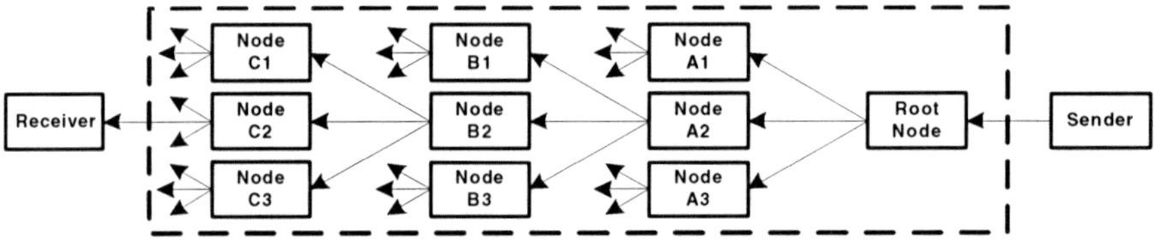

Fig. 5.8. Topology for the advanced hash-based infrastructure

Just as in the other approaches, the entry point of the infrastructure is the root node. Between this node and the receivers, there are several levels of hierarchy of intermediate nodes. In contrast to the basic hash-based approach, now there exist several paths between the root node and a certain receiver. In figure 5.8, a path to a receiver is shown that traverses the nodes A_2–B_2–C_2. Alternative paths could be

A_2–B_1–C_2 or A_1–B_2–C_2 or A_1–B_1–C_2 if C_2 had both B_1 and B_2 as parent nodes and B_1 and B_2 each had both A_1 and A_2 as parent nodes. As a single intermediate node can now have many parent nodes and an infrastructure depth of at least $t = 5$ makes sense, the number of different paths to a different receiver increases considerably. This provides much more variability compared to the basic hash-based approach.

Each of the nodes in the infrastructure assigns node identifiers to its child nodes. They are similar to the ones in the basic hash-based approach, but there can be more than one for each child node of a node. In the following, N is used to denote a child node and A (for "aliases") is used to denote the different node identifiers of such a node. P (for "parents") is used to denote the parent nodes of a node. A node only accepts messages from its respective parent nodes.

The intermediate nodes can be in two modes of operation. On levels near the root node, the random numbers already known from the basic hash-based approach are required to make pseudonyms denoting the same path able to appear different. Using the random numbers to obtain variation is the first mode of operation. The random numbers are not required on deeper levels of hierarchy since these levels no longer have the goal to prevent tracking but to conceal the receiver. Therewith, the second mode of operation is a simpler one for deeper levels of hierarchy. It does not use random numbers for variation purposes.

The root node itself and the nodes on at least the first levels below it use the first mode of operation so that tracking can be prevented. This mode is similar to the one in the basic hash-based approach but makes use of the node identifier aliases and additional private values for protecting the infrastructure.

Each such node has a number of database tables. One database table links the node identifiers N_i of its child nodes to the hostnames of the child nodes as already known from the basic approach. A new database table lists the node's parent nodes and assigns each parent node a private random value v:

$$P \longrightarrow \text{parent node}, v \quad \forall P$$

In another database table, each node alias A maps to the node N it belongs to and a private random value u:

$$A \longrightarrow N, u \quad \forall A$$

A fourth database table exists that is the same as in the basic hash-based approach. In this table, the node identifiers are not used directly. Instead, the node identifier aliases A are used:

$$h(k, A) \longrightarrow k, A \quad \forall A, k \in [0, r_{\max}[\subset \mathbb{N}_0$$

h again denotes the one-way hash function, and the parameter $r_{\max}$ defines a trade-off between the number of different pseudonyms that can be created for a particular receiver and the space that is required in the database for this purpose. For instance, if $r_{\max} = 2^8$ and this mode of operation is used on the first $t = 3$ levels of hierarchy, ${r_{\max}}^t = (2^8)^3 = 2^{24} = 16777216$ different pseudonyms can be created for a particular path when using a particular set of node identifier aliases. r_{max} does not need to be the same on all levels and can thus be adjusted to the requirements individually. In and near the root node, it should preferably be higher as on lower levels of hierarchy to get a better protection against tracking by constellation.

Nodes near the receiver that operate in the second mode of operation do not require the database table with the hash values. Instead, a higher number of node identifier aliases can be stored while keeping the size of the database constant. The other tables are the same as the ones in the first mode of operation.

Path provisioning and pseudonym generation

When a receiver joins the infrastructure, a path to the root node and – corresponding to it – a set of node identifier aliases of the intermediate nodes need to be determined. For this to be done, the receiver contacts a node on the lowest level of hierarchy of the infrastructure and joins as child node. The node selects one of its parent nodes P which itself selects a node identifier alias A of the node. The same is done on all the levels further up until the root node is reached: Each respective node selects a parent node (if the root node is reached there is no choice any more), and this parent node selects a node identifier alias for the respective node.

For nodes operating in the second mode of operation in which no random number based variation is used, not the node identifier aliases themselves are given away. Instead, they are scrambled using applications of the function s along the way through the nodes operating in this mode of operation. As parameters, the private values v that correspond to the respective selected parent nodes and the private values u that correspond to the respective selected node identifier alias are used. This makes the information that is later on given away for pseudonym generation be dependent on the selected path. The scrambled node identifier aliases are denoted by A^* in the following.

Nodes in the first mode of operation calculate values w using a function that uses the private values u and v as input: $w_i = w(u_i, v_{i+1})$. The function w needs to have special characteristics that will be discussed later on.

To be able to generate pseudonyms, the node identifier aliases A_i and the values w_i are required for the nodes operating in the first mode; and for the nodes operating in the second mode, the scrambled node identifier aliases A_i^* are required.

Pseudonym generation is similar to the basic hash-based approach but also uses the new values w_i for an additional, path dependent scrambling. For this scrambling, an invertible function g is used with the values w_i as parameter. For a straightforward implementation, the inversion of g should be possible in two independent steps using the input variables of the function w. This means that if $X^* = g_{w_i}(X) = g(X, w_i)$ then $X = g^{-1}(X^*, w_i) = g^{-1}(X^*, w(u_i, v_{i+1})) = g^{-1}(g^{-1}(E, u_i), v_{i+1}))$ should hold. The simplest implementation would be to use the xor function for the functions g and w, so that $w_i = u_i \oplus v_{i+1}$ and $X^* = g(X, w_i) = X \oplus w_i$ and $X = g^{-1}(X^*, w_i) = g^{-1}(X^*, w(u_i, v_{i+1})) = X^* \oplus (u_i \oplus v_{i+1}) = g^{-1}(g^{-1}(E, u_i), v_{i+1})) = (E \oplus u_i) \oplus v_{i+1}$. The stated requirement is fulfilled here because of the xor function being associative.

For instance, if the root node and the nodes on the next two levels used the first mode of operation and further nodes on the next two levels used the second one, a pseudonym vector would be calculated using the following calculation rule:

$$P_0 = \begin{pmatrix} h(A_1, r_0) \\ f_{r_0}(g_{w_0}(h(A_2, r_1))) \\ f_{r_0}(g_{w_0}(f_{r_1}(g_{w_1}(h(A_3, r_2))))) \\ f_{r_0}(g_{w_0}(f_{r_1}(g_{w_1}(f_{r_2}(g_{w_2}(A_4^*)))))) \\ f_{r_0}(g_{w_0}(f_{r_1}(g_{w_1}(f_{r_2}(g_{w_2}(A_5^*)))))) \end{pmatrix}$$

in which $A_4^* = s(A_4, v_3)$ and $A_5^* = s(s(A_5, v_4), u_3, v_3)$ holds. The sender needs to perform the same operations for the fourth and the fifth element of the pseudonym vector because it can directly use the already scrambled values A_4^* and A_5^*.

Like in the basic hash-based scheme, r_i is a random number within the valid range. The function f and the hash function h are also the same as in the basic scheme and will thus not be explained again. The length of each element p_i of the pseudonym vector has the length of the output of the employed hash function, e.g. 128 bit or 160 bit. All other functions are one-to-one mappings for which domain and co-domain have the same number of bits.

Message forwarding

A message with a pseudonym vector P for defining the receiver is first sent to the root node of the pseudonymization infrastructure. See figure 5.9 for an example in an RFID scenario. The element p_0 of the pseudonym vector denotes a child node of the root node on the path to the receiver that is defined by the pseudonym vector and also includes the random number r_0 used in the generation process. p_0 is treated analogously to the basic hash-based scheme: The root node looks up $p_0 = h(A_1, r_0)$ in the database table to get the preimages A_1 and r_0. A_1 denotes the child node to which the message will be sent next. Before this is done, r_0 is used to invert the outer functions f of the other elements in the pseudonym vector. In contrast to the basic hash-based scheme, A_1 is a node identifier alias to which in the database table a randomly generated private number u_0 has been assigned. This number u_0 is used to partly invert the outer function g of the other elements of the pseudonym vector. Afterwards, the used element p_0 is stripped off the pseudonym vector, and the message is sent to the next node which is specified by the information that is stored in a database table record that is linked to the node identifier alias A_1.

In the next node, the inversion of the outer functions g of the pseudonym vector elements is completed using the value v_1 that is obtained from the database table of valid parent nodes of the node. Now the structure of the pseudonym vector corresponds to the initial one that was sent to the root node.

In the current node and all the subsequent nodes that operate in the first mode of operation (the one with the random values r_i for variation), the actions that were taken by the root node are performed analogously.

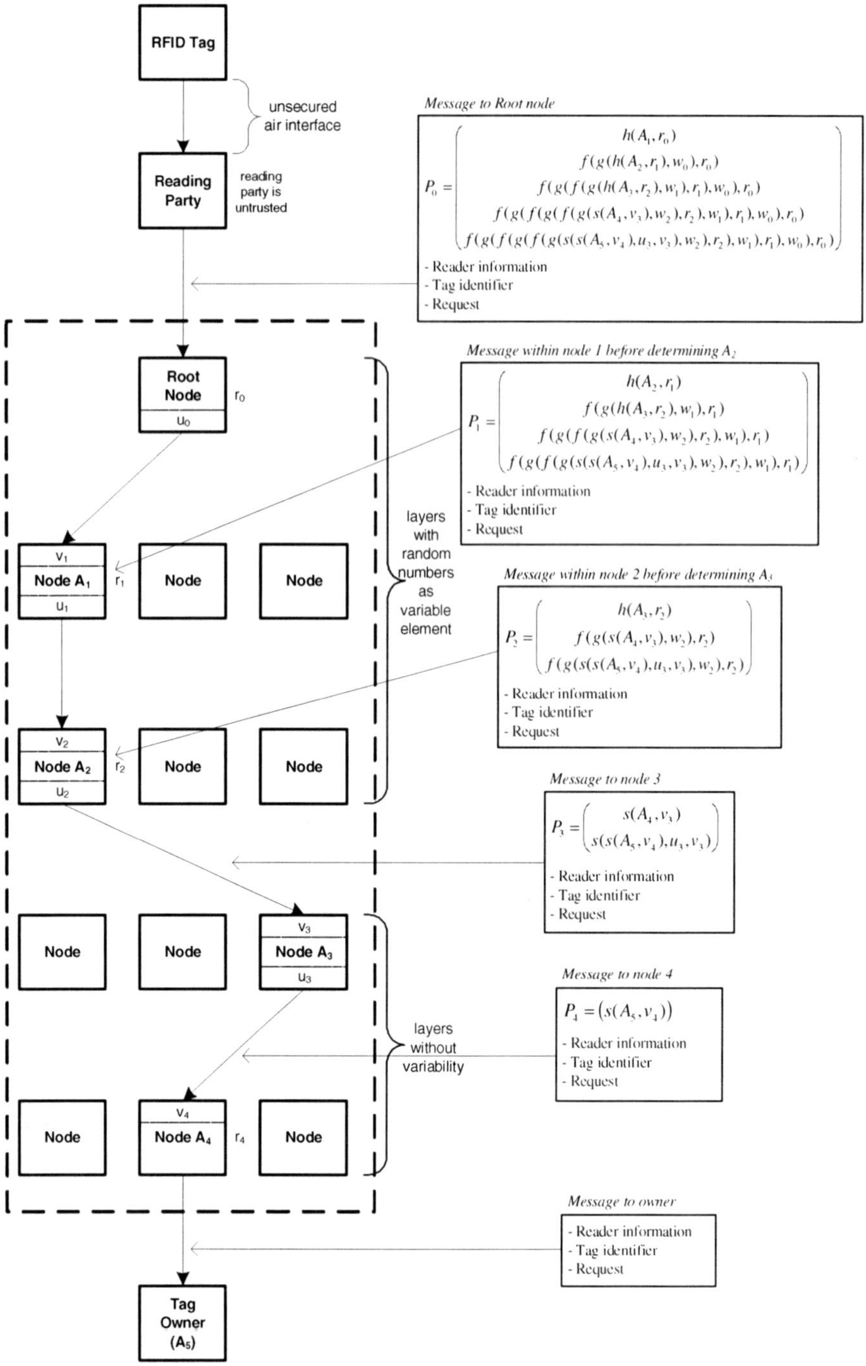

Fig. 5.9. Example of the advanced hash-based approach in an RFID scenario

For nodes using the second mode of operation, in which no random number based variation takes place, the process in the nodes is simpler: The values u and v that are stored in association to the involved parent and child nodes of a node are used to invert the outer functions s of the remaining elements of the pseudonym vector. This reveals the node identifier alias of the next node to which the message will be sent. The corresponding node information is determined using the database tables, and the no longer needed first element of the pseudonym vector is stripped off. Then the message is sent to the next node. These steps are performed for each remaining node on the path until the message ultimately reaches the receiver whereby the last element of the pseudonym vector is gone.

Enhancements

The functions f, g, and s should be chosen in such a way that the order of their applications with different parameters should not be interchangeable, i.e. applications of the functions should be non-commutative. This might give additional security in case intermediate nodes are under control of attackers.

In addition, the enhancements stated for the basic hash-based approach are also applicable here.

Discussion

The presented approach tries to address the weaknesses found in the basic hash-based scheme. It introduces additional variability by allowing different paths to the same receiver whereby the paths themselves can be represented by different node identifier aliases. To address the different roles of the nodes on different levels of hierarchy, operation of the nodes is split into two modes. Additional randomly generated private values u and v are used to bind the elements of the pseudonym vector to the particular path.

RELIABILITY: Reliability of the infrastructure using the presented advanced hash-based approach mostly equals the basic one (see section 5.7) so that it needs not be discussed here again. The only difference is that a receiver does not need to join the infrastructure twice to get an alternate path through the infrastructure and therewith a backup path.

SECURITY: Like the other approaches, the advanced hash-based scheme builds upon the principle of shared trust by using pseudonyms that only multiple nodes together are able to resolve. It works similar to the basic hash-based approach but has the goal to eliminate the discovered weaknesses.

Compared to the basic hash-based approach, the advanced one introduces the node identifier aliases and therewith provides more variability and a more flexible topology enabling different paths to each receiver. Additionally, the additional private values u and v are added to scramble the elements of the pseudonym vector dependent on the chosen path through the pseudonymization infrastructure. Thus, it is no longer possible to reconstruct the complete infrastructure with all databases just by knowing the node identifiers in the system and the mapping between the identifier of the receiver and its real network address.

The pseudonyms are composed the same way as in the basic hash-based approach: a pseudonym vector with one element for each node to be traversed. If an attacker is able to collect one or many of such pseudonyms, he is not able to reveal any information about the corresponding receiver without additional information, i.e. the data stored in the database tables of the intermediate nodes. This means for the RFID scenario that an attacker would not be able to recognize the entity that is in charge of a queried tag.

The random numbers that are used in nodes operating in the first mode of operation to enable the generation of different pseudonyms using the same node identifier aliases prevent unwanted tracking. Due to the one-way characteristic of the hash function, without additional information, an attacker is not able to link different pseudonyms describing the same path and thus denoting the same receiver. Moreover, he is not able to create pseudonyms by himself as long as he is not in possession of valid node identifier aliases and the corresponding private values u and v.

The number of different pseudonyms that can be created for each path described by a particular set of node identifier aliases depends on the value $r_{\max}$ of the nodes using the first mode of operation. For instance, if there are three levels in that mode and $r_{\max}$ is the same on all these levels and has a value of $2^{10} = 1024$, there can be ${r_{\max}}^3 = 2^{30}$ possible pseudonyms. Note that all these pseudonym vectors are not completely different from each other. For instance, there are only $r_{\max}$ different first elements of the pseudonym vector. Therefore, $r_{\max}$, especially for the root node, may not be decreased too much to effectively prevent tracking and tracking by constellation.

Thus, like in the basic hash-based scheme, the advanced approach provides an effective protection against attackers between the sender and the root node. In case of RFID systems, this means that the insecure wireless link between tag and reader as well as the reading entity which might be malicious is protected.

The advanced hash-based approach intends to solve the weaknesses of the basic one which are mainly the static topology and the few static node identifiers. In the advanced approach, the topology is much more flexible and enables different paths to the same receiver. The few node identifiers are superseded by a much higher number of node identifier aliases. The possibility to combine different paths and different node identifier aliases gives much more variability compared to the basic hash-based scheme. The additional private values u and v that are linked to the node identifier aliases of a node's child nodes and a node's parent nodes, respectively, introduce additional secrets into the infrastructure so that it cannot be attacked as easily as the basic hash-based one by collecting node identifiers and rebuilding the database tables of the nodes.

Thus, the protection against collecting data about the pseudonymization infrastructure is much better compared to the one of the basic hash-based approach. Nevertheless, the advanced approach still has the problem that node identifier aliases and values w that are calculated using values u and v have to be given to the sender. This becomes a problem if an attacker harvests and tries to abuse these data: The number of node identifier aliases that are stored in the root node is limited. An attacker that has collected a high number of node identifier aliases on that level, e.g. by physical extraction out of a tag, can rebuild at least parts of the root node's database tables. Therewith, he would be able to obtain the second level node identifier aliases if a pseudonym was given and he would thus be able to perform tracking by constellation. The information needed for pseudonym generation should thus be kept as private as possible. In the RFID scenario, the entities in charge of tags do not require the information needed to create pseudonyms and thus should not get it. It would be sufficient to store the information in the tags so that it cannot be obtained out of them by means different from physical extraction. But in contrast to the basic hash-based scheme, the high number of node identifier aliases, the additional secrets, and the higher flexibility in the topology make it much more difficult for an attacker to collect and reconstruct the information needed to reveal the identity of the receiver. Here, the security characteristics of the advanced hash-based approach are again much better than the ones of the basic approach.

If an attacker is able to obtain the information that is required for creating pseudonyms, i.e. physical extraction out of a tag in the RFID scenario, then he is able to generate a list of valid pseudonyms. With such a list, he is able to link the pseudonyms created by this sender. Therewith, the attacker becomes capable to track the sender. This attack is the same as the one in the basic hash-based scheme. But as there are multiple paths to a receiver in the advanced hash-based approach and there are different sets of node identifier aliases describing a certain path, an attacker is no longer able to link all pseudonyms denoting the same receiver. This means for the RFID scenario that even if an attacker becomes able to track a single tag, he is not able to track all the tags belonging to the same receiver, i.e. the entity in charge of the tag. This is not the case in the basic hash-based scheme. Nevertheless, both hash-based schemes are weaker than the one using asymmetric cryptography because having the information to generate pseudonyms does not enable linking any pseudonyms there in moderate time-memory-complexity (generating all possible pseudonyms that can be created would be the only attack besides breaking the asymmetric cryptography scheme itself).

RESOURCES: The resource consumption is generally quite similar to the one in the basic hash-based approach so that only the differences will be discussed in the following.

The application of the additional function g which uses the values u and v as parameters do not impose high a computational burden on sender and intermediate nodes compared to the hash calculations. Since for intermediate nodes operating in the second mode of operation the sender does not need to calculate a hash function when generating the corresponding element in the pseudonym vector, the advanced approach needs less computation here.

SCALABILITY: Scalability characteristics of the presented advanced approach are the same as those of the basic approach. The only difference is that the topology is no longer bound to a tree: Adding additional nodes to the infrastructure is possible, either as mirrors to existing nodes or as new independent nodes.

5.9 Hash Collisions and Pseudonym Shortening in Hash-Based Approaches

Using the hash-based approaches, each node has a database table that maps hash values $h(\ldots)$ to the corresponding preimages, i.e. the random number used for varying the pseudonyms and the node identifier or node identifier alias, respectively. This mapping needs to be non-ambiguous to enable a node to determine to which unique node to send a message next. This is only ensured with a certain probability. This will be analyzed in the following.

Hash collisions

The probability of hash collisions in general was already discussed in section 2.4.3. With standard hash functions, we have a domain d for the hash values of 128 bit or 160 bit. In the following, we assume $d = 128$. In the basic hash-based approach, there could be for example $r_{\max} = 2^{16}$ and $\#N = 2^8$ child nodes of node, which would result in $\#records = \#N \cdot r_{\max} = 2^{24}$ records stored in the database table with the hash values. We therefore assume that the number of records is $c = 2^{24}$. For the advanced hash-based approach, the number of records is similar because $r_{\max}$ will usually be smaller and the number of node identifiers due to the use of node identifier aliases higher.

Using the already presented approximation formula from Sayrafiezadeh [Say94], we obtain with the given values:

$$P_{\text{coll}} = 1 - \left(1 - \frac{c}{2d}\right)^{c-1} + \varepsilon = 1 - \left(1 - \frac{2^{24}}{2 \cdot 2^{128}}\right)^{2^{24}-1} + \varepsilon \approx 4.1 \cdot 10^{-25}$$

with a negligibly low error of

$$\varepsilon < \frac{c^3}{6(d-c+1)^2} = \frac{(2^{24})^3}{6(2^{128} - 2^{24} + 1)^2} \approx 6.8 \cdot 10^{-57}$$

Thus, the probability of hash collisions when creating the database table for the hash values is very low. If nevertheless a collision occurs, one has three options. The first: Store both possibilities in the database table and treat and route the message more than once. One of the messages will take the intended path, the other one/ones will with very high possibility lead to the error that the receiver is unknown. The second: Store none of the possibilities which will lead to the error that the receiver is unknown. The third: Store one of the possibilities and discard the others which will either lead to a correctly routed message or with very high probability ($P_{\text{invalid}} = 1 - (\frac{2^c}{2^d})^{t_r}$ in which t_r is the number of remaining levels through which the message still needs to be routed) the error that the receiver is unknown. Due to the expectations of the receiver and the explanatory power of the error message, the third option is recommended. But note that all these options only become relevant when the respective colliding entries are accessed. This happens only if in a pseudonym a node identifier (or the node identifier alias in the advanced hash-based scheme) and a random number for pseudonym variation are used that belong to a colliding hash value.

Pseudonym shortening

As calculated in the previous section, the probability of hash collisions in the database tables is negligibly low. This is due to the high number d of bits that each hash value calculated using a standard hash function has. But this high number of bits has also disadvantages. As already stated in the description of the hash-based approach (see section 5.7), the size of pseudonym vector elements equals the size of the hash values. A complete pseudonym vector that has t elements is still much shorter than a pseudonym using the asymmetric cryptography approach but nevertheless becomes quite big, i.e. 80 byte if $t = 5$ and hash values have 128 bit.

One can reduce the size of the pseudonym vector by shortening the individual pseudonym vector elements. The scrambled hash values in these elements are only used as an index within a database table of a node. As this index is longer than required for uniquely identifying a single record with a high probability, one can shorten the hash values and therewith the pseudonym vector elements by only using the first d_{short} bits of a hash value. This makes the pseudonyms smaller and therewith reduces the amount of data that a sender needs to transmit. Moreover, the smaller hash values save on space in the database. On the other hand, by shortening the hash values the probability for hash collisions is increased. Therefore, a suitable trade-off between the size of the hash values and the probability for hash collisions resulting in potentially undeliverable messages needs to be found.

Such a trade-off could be to use hash values with the double number of bits that is required to represent all records in the database tables with the hash values. If, for instance, $c = \#records = 2^{24}$ records are stored in the database that require $d_{rec} =$ 24 bit for representation, one could truncate the hash values to $d_{\text{short}} = 2 \cdot d_{rec} = 48$ bit. The approximated probability that hash collisions occur in the database table would be for these values:

$$P_{\text{coll}} = 1 - \left(1 - \frac{c}{2d}\right)^{c-1} + \varepsilon = 1 - \left(1 - \frac{2^{24}}{2 \cdot 2^{48}}\right)^{2^{24}-1} + \varepsilon \approx 39.3\%$$

with a negligibly low error of

$$\varepsilon < \frac{c^3}{6(d - c + 1)^2} = \frac{(2^{24})^3}{6(2^{48} - 2^{24} + 1)^2} \approx 9.9 \cdot 10^{-9}$$

The probability for hash collisions looks high, but to use the double number of bits for the hash values is nevertheless a very conservative approach: For all the

$\#records = 2^{24}$ records in the database table in $P_{\text{nocoll}} = 1 - P_{\text{coll}} \approx 60.7\%$ no collision at all occurs. As the number of collisions is binomially distributed, the probability that one uses some of the relatively few hash preimages out of the 2^{24} valid ones that result in a collision is very small.

In sum, it makes sense to truncate the hash values used in the pseudonymization infrastructure to save storage space and decrease the size of the pseudonym vectors. In an infrastructure in which 128-bit hash functions are used and in which the exemplary numbers stated above are used, the sizes decrease by 62.5% with a for practice negligibly increase of the probability that a message cannot be routed due to hash collisions.

5.10 Summary and Research Directions

Unstructured identifiers do not contain information about the entity in charge of an RFID tag because for privacy reasons, the identity of that entity must not be revealed. This becomes a problem when building large, inter-organizational RFID systems. In such systems, a means is required to query the responsible organization for additional information about a tag without revealing this organization's identity. In this chapter, the problem has been described, the use of pseudonymization infrastructures for solving this problem has been proposed, and different implementation concepts have been presented.

First, a pseudonymization infrastructure based on asymmetric cryptography has been discussed as possible solution. This concept has advantageous characteristics: The approach is straightforward, well scalable, and secure. If protection of location privacy is an issue in the RFID scenario, much computation needs to be performed by the tags which is well beyond the capabilities of low-cost tags and would take high a processing time even for much more expensive tags. Furthermore, due to the blocksize of the used ciphers, the pseudonyms become quite large.

Consequently, a more lightweight concept that is better tailored to the demand in RFID systems is favored. Therefore, a new concept based on one-way hash functions as basic primitive has been introduced. Starting with a basic approach, the concept has been improved to an advanced approach with better security characteristics. Security in the hash-based concept is based on the one-way property of the employed hash function and the data security of the databases in intermediate nodes. Analogously to the concept using asymmetric cryptography, messages are processed and forwarded by intermediate nodes following the principle of shared trust.

The hash-based concept is less resource consuming than the concept based on asymmetric cryptography and has similar characteristics concerning reliability and scalability. For instance, adding receivers does not require a change in any intermediate node beyond the leaf node of the infrastructure where the receiver joins. Furthermore, pseudonyms can be made much shorter than in the concept based on asymmetric cryptography while still providing a high number of different pseudonyms for each receiver. Optimization of the pseudonym size has been discussed in a separate section.

On the other hand, the security characteristics of the hash-based concept are not as good as the ones of the approach using asymmetric cryptography for particular attacks: Having the required data for creating pseudonyms enables tracking of the respective tag. But this is usually not a problem in the RFID scenario because physical extraction would be required to obtain that data. A more serious problem is that even in the advanced hash-based approach, security cannot be assured in the long term: Data collection still enables to perform tracking by constellation. Nevertheless, even if the newly introduced hash-based concept is not mature enough for practical application and thus more research is required, a large step towards less resource consuming pseudonymization infrastructures with similar favorable characteristics compared to infrastructures based on asymmetric cryptography was done.

Independently of the concept the pseudonymization infrastructure is based on, there are some issues for practical application of such infrastructures in the RFID scenario: To implement the concept of shared trust, the messages need to be routed through and be processed by several intermediate nodes. This introduces delay that is a problem for interactive applications that expect the result of a tag query within a very short time. Another problem is that the request messages of the reading entity in the RFID scenario are sent in clear to the root node. For instance, an intelligence service that keeps the root node and its mirrors under surveillance can eavesdrop and obtain interesting traffic patterns as well as information about the readers and their purposes. This problem can be easily addressed if the asymmetric cryptography based approach is used and the tag includes the reading entity's request in the routing onion. A third problem is that it is difficult to assure the security of the pseudonymization infrastructures in the long term. The infrastructures rely on private keys or similar information. These may get compromised. In such a case, there is no possibility to update the whole system with new key material because the public keys (or node identifiers or whatever material is used) that are stored in the tags are not regularly accessible since no permanent connection to the Internet exists. All the stated issues should be put into consideration before pseudonymization infrastructures are used in real-world RFID applications.

6

Extending the RFID System Model

This chapter deals with building inter-organizational RFID systems that respect the different interests of the involved entities regarding functionality, security, and privacy. Some of the underlying ideas have already been part of the discussion in [HMM04]. They will be presented again here in a more concrete and structured manner and will also be extended.

In chapter 4, it has been shown that message based modification needs to be performed on every usual tag query to protect location privacy effectively. This required the communication of ordinary readers with a backend entity in charge of the tag. As no backend identifier could be included in the tag identifier because that would allow an attacker to perform unwanted recognition and tracking by that backend identifier, a central backend entity had to be used.

In the previous chapter, pseudonymization infrastructures have been introduced to solve the problem to contact a responsible backend entity without revealing that entity to a reader who is a potential attacker. This new functionality will be integrated into an overall RFID system model in this chapter.

The model will be extended further to enable building of truly open, inter-organizational RFID systems using a push model that includes all involved entities that are interested in obtaining data and data processing. The looming possibilities go much beyond the possibilities that are present using just a global numbering scheme like the EPC and the classic RFID system model.

In the next section, the classic RFID system model is presented. Afterwards, the extension that has been introduced in the previous chapter is integrated into the model, and the push principle for RFID systems is brought into discussion. The successive section introduces the tag bearer as separate entity because he is the user that is mainly concerned regarding privacy protection. Based on the proposed entities, a "personal manager" is considered to manage data, context, security and privacy policies and the like for an entity. Lastly, the complete RFID model with the building blocks introduced in the previous two chapters and in the current chapter is assembled and rated to derive the research directions for the next chapter.

6.1 Classic RFID Model

Most literature regarding security and privacy in RFID systems is (at least implicitly) based on the following three entities or groups of entities:

- RFID tags,
- legitimate entities and their equipment, and
- other entities including attackers and their equipment.

The first group of entities is the RFID tags. These tags are an indispensable part of the RFID system so that they play a crucial role in all RFID system models.

The second group of entities relevant for RFID protocols regarding security and privacy are legitimate entities, i.e. one or more entities that are in charge of the tags and that are allowed to control them, e.g. to perform identifier modification. These entities operate different system parts like readers and backend databases. The important aspect is that all parts of the equipment of the legitimate entities are trusted from a system point of view.

In contrast to the second group of entities, the third group includes all other entities and is made up of all other equipment. This equipment is not trusted and in theory does not have the ability to control the tags. As parts of that equipment might be operated by attackers, access to the rest of the RFID system, i.e. tags and legitimate equipment, is limited and strictly controlled.

Inter-organizational use of the model

RFID protocols proposed in the literature that aim at ensuring security and privacy operate between the tags and the respective legitimate entity. The third presented entity does not have the information like shared secrets that are only known to tags and legitimate entity. Thus, this third entity is not able to control the tag. This is expressed by Sarma et al. in [SWE02]: "Both tags and [legitimate] readers should trust each other." The protocols aim solely at securing the communication via the wireless air channel and do not consider the infrastructure and its organization beyond the readers. Another example is presented by Juels and Pappu in [JP03]. They describe an approach based on external re-encryption that requires trusted readers.

The problem with the classic RFID model is that it is based on the requirements that stem from the RFID protocols. The practical requirements are not taken into account in the system organization which gets eminent if inter-organizational RFID systems are considered. This procedure is not adequate: Today's supply chains consist of a number of actors, i.e. many different organizations that work together within a process. One of these organizations might be the issuer of the tag and shall have full control over it, other organizations, e.g. subcontractors, shall be able to read it, other organizations not. These different roles result in different levels of trust so that simply distinguishing trusted readers and untrusted readers is escapist.

If different organizations shall be able to read out the tags and RFID protocols aiming at security and privacy are used, the organizations need to be trusted and get the required shared keys. With these keys, all the organizations gain control over the tags which is often not desired. This limits the applicability of the classic model and of systems adhering to it.

The data sharing requirement

Besides the stated trust problems in inter-organizational RFID systems, data sharing between multiple organizations is of huge concern today. Under the assumption that the organizations are trusted and thus have got the key material to identify the tags, the organizations process the tag reads and store the data in their backend, see figure 6.1. In practice, the data is processed by RFID middleware and then stored and used in enterprise resource planning applications or the like.

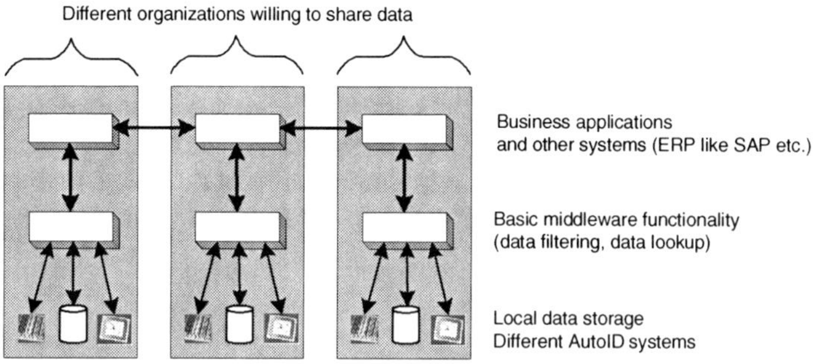

Fig. 6.1. Data sharing in the classic RFID model

If data regarding tag reads shall be shared among different organizations, the data is usually shared on application level via defined interfaces. There are no open standards regarding required characteristics of that data sharing interface, e.g. regarding access control. The interfaces can thus be regarded application specific so that the effort required for sharing data between many organizations gets high.

6.2 Untrusted Reading Entities

As the classic RFID model proves to be inappropriate, it is adapted in this section. Still, three groups of entities are distinguished:

- RFID tags,
- legitimate entity, i.e. tag owner, and
- reading entities operating readers.

At first sight, there is not a huge difference, but it exists: In this new model, readers and the entities operating these readers are untrusted by default. There are no trusted readers at all in this model. This is completely different to the classic model.

Still, there are the RFID tags. But now there is a single legitimate entity that is in charge of a respective tag. Beyond that, there are arbitrary reading entities. Whether these entities are trusted or not or even trusted regarding some aspects only, is not defined in the model. The default, however, is that the reading entities are untrusted. Legitimate readers need to authenticate to the legitimate entity before they are regarded trusted. Note that the legitimate entity can operate readers on its own. In that case, legitimate entity and reading entity coincide.

Message based identifier modification in the context of the new model

Let us consider an example for demonstration. In chapter 4, the "Hash-based ID variation" approach has been introduced. It is a protocol that performes identifier modification based on a message exchange. For such solutions to provide location privacy, the protocol needs to be run regularly, i.e. ideally on every tag query. This means that all readers should take part in using the protocol whether these readers are operated by the legitimate entity, a trusted organization, or by anybody else that does not aim at disturbing proper system operation.

This requirement directly leads to the model presented in this section: There are the RFID tags, and there is a central legitimate entity. Third, there are readers that might be operated by nearly anybody but that are not trusted by default.

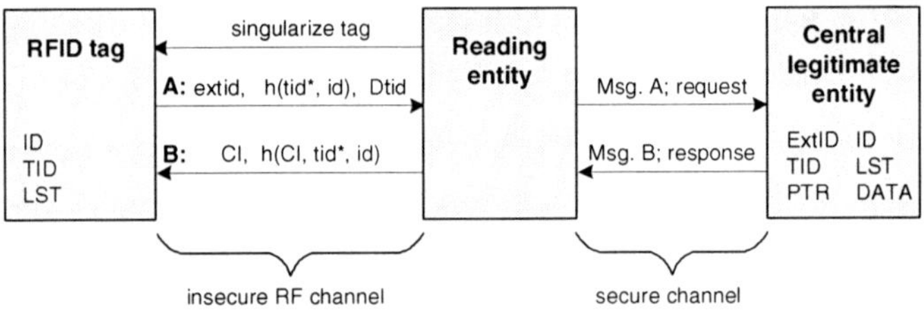

Fig. 6.2. Hash-based ID variation and untrusted readers

Within the protocol, a reader queries the tag and receives message A, see figure 6.2. A reading entity is not able to identify the tag because the identifier appears as a random number. Thus, the message is forwarded to the central legitimate entity along with a request asking for tag data. The legitimate entity is able to identify the tag using the information in its tag database and sends message B and the response to the reader's request back to the reading entity. The reading entity finally forwards the message B to the tag so that the identifier modification can be performed successfully.

In this process, the reading entity does not require and also does not obtain any useful information from the protocol messages. Thus, no trust is required regarding the reading entity. It is just presumed that all reading entities apart from attackers do not misbehave and complete the protocol. The legitimate entity can decide whether to reveal the requested tag data to the reading entity or not. A fine-granular access control policy regarding tag data can be implemented here.

In chapter 5, pseudonymization infrastructures have been used so that the legitimate entity does no longer need to be central. In this scenario, the reader sends the message A to the pseudonymization infrastructure that forwards the message to the right legitimate entity in charge of the respective tag.

Push principle

The described procedure has interesting consequences: It does not matter whether the legitimate entity operates readers by itself or whether others operate the readers. All readers in the RFID system query tags, and the result is forwarded to the respective legitimate entity in charge of the tag.

The consequence is that an organization needs not operate readers on its own. Already existing readers of other organizations or from public infrastructure also provide data regarding tag reads. As a slogan, one could thus say "Every reader is also your reader" to circumscribe this.

Applications thus get much more fine-grained data: The density of readers is much higher when each organization does not need to operate readers on its own but all readers can be used as shared infrastructure.

Regarding privacy, the new possibilities that enable a better, more fine-grained tracking of goods appear counterproductive because this also provides more possibilities for unwanted tracking. But the positive thing is that the push principle makes the potential for abuse that also exists in classic systems more obvious. The increased transparency caused by the clear data paths in the system thus provides an advantage for consumers.

One should also note that the potential for abuse caused by such a push scenario is not new but just not that evident in the classic RFID system design. Using the Object Naming System (ONS) proposed by EPCglobal/GS1, requests regarding additional information regarding an RFID tag are forwarded to the manufacturer of the product or to another responsible entity. Thus, the latter entity also gets detailed information regarding tag reads albeit these reads are not performed by readers operated by that entity. Thus, there is an implicit push scenario.

In sum, the push principle enables truly open, inter-organizational RFID systems providing a variety of new possibilities. The effective inter-organizational use of reader hardware provides economic advantages and more detailed tracking data. The requirement to implement powerful privacy protection measures in RFID systems becomes evident in such a scenario and will be addressed in the next section.

Inter-organizational systems and data sharing

Having untrusted reading entities eases the creation of inter-organizational RFID systems. With the presented push principle, the legitimate entity identifies the tag and is the only entity that needs to have the key material that is required for controlling the tag.

All readers are regarded illegitimate and are untrusted at first. Then the entity that is in charge of the tag, e.g. the owning organization, decides whether a read request is legitimate or not and therewith whether data regarding the tag is revealed to the reading organization or not. This enables flexible access control schemes for controlling access to tag information and enables a powerful logging of the divulgement of data if required.

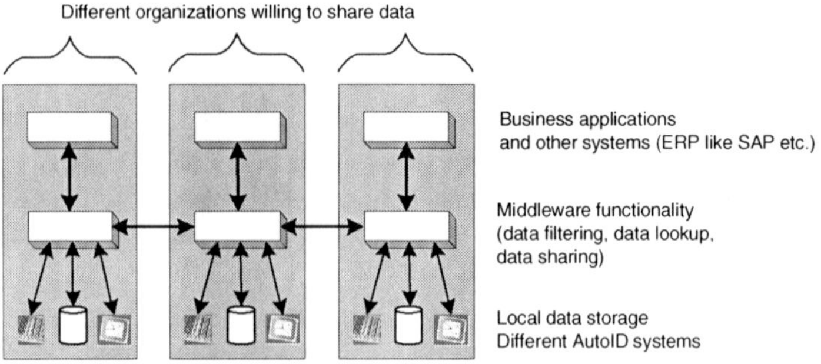

Fig. 6.3. Data sharing in an extended RFID model

The push principle eliminates the need for sharing data between organizations on application level: The legitimate entity of a tag returns data to the reading entity using an interface defined on RFID system level if the divulgement of data is appropriate. In addition, the legitimate entity as well as the reading entity can forward the data to other organizations that require the data. As shown in figure 6.3, data sharing is thus done as middleware functionality from the RFID system. The data sharing is therewith transparent to applications so that from an interface point of view it makes no difference any more whether the tag query has been performed by readers of the own organization or by readers from other organizations that shared their data.

Imagine for example a logistics service company. When the company receives the item to be shipped, it attaches an RFID tag to that item. This way, the item can be identified automatically if required. When the tag is queried by a reader of the logistics service company, the reader forwards the data to the legitimate entity which is the logistics company itself. The company can use the data for tracking the item in its logistic network by making the data available to tracking applications. For instance, information regarding the shipping status can be made available to the customer via the company website. The described proceeding is the standard one and can be realized with the classic RFID model in the same manner.

A more interesting case is when a subcontractor is assigned the task to deliver the item on parts of its way. In this case, the subcontractor shall be able to identify the item using the RFID tag that has been affixed by the logistics company. This is very simple using the new model: When the subcontractor reads the tag, he is not able to identify it because the reader is regarded untrusted. Thus, the subcontractor needs to forward the response from the tag to the legitimate entity, i.e. the logistics company who issued the tag because only this entity is able to identify the tag. Now, the legitimate entity can decide using a policy whether or not to reveal the tag identity and optionally additional tag information to the reading entity, i.e. the subcontractor. As the subcontractor is entrusted by the logistics company, the tag data is revealed so that the subcontractor gets the required information. If the subcontractor reveals additional information regarding the tag read, e.g. the location of the reader, the logistics company can update the shipping status that is made available to the customer accordingly.

Each entity involved can also forward data regarding tag reads to business partners. Instead of providing shipping information by the logistics company on the website, the logistics company could also forward data regarding tag reads to the respective customer who can use it in his own ERP systems.

An alternative is that the customer has already attached the RFID tag to the item to be shipped. In this case, the customer is the owner of the tag and thus its legitimate entity. When the logistics company reads out the tag, it needs to ask the legitimate entity for the tag identification. As the logistics company is contracted by the customer, it will get the required data and can use it in the same way as if the logistics company is the tag owner.

The example shows that the push concept using readers that are untrusted at first eases inter-organizational data sharing. The system becomes truly open. The readers becoming shared equipment and the generic data sharing concept results is immense economic advantages and increased transparency compared to the classic RFID system model. Note that the push concept and the inter-organizational data sharing on the middleware layer can also be used with other auto-id systems like barcodes.

6.3 Tag Bearer as Additional Entity

With the RFID system model presented in the previous section, the entities are defined as shown in the upper part of figure 6.4: There are the tags, the reading entities, and the tag owners in charge of the tags. In addition, there might be a pseudonymization infrastructure between reading entities and tag owners as introduced in chapter 5.

The model is completely sufficient for the operation of RFID systems. The push model even exceeds the possibilities that classic systems pose. But from the perspective of data security and privacy, the model does still not adhere to the practical demands. Consumers want to be in control, otherwise they oppose the technology [GS05].

Imagine the use of RFID tags for subway ticketing. The tickets are then equipped with RFID tags. The operating company of the transport system who issues the tickets is the owner of the tags. With readers at the subway station, the system is complete and can serve its purpose.

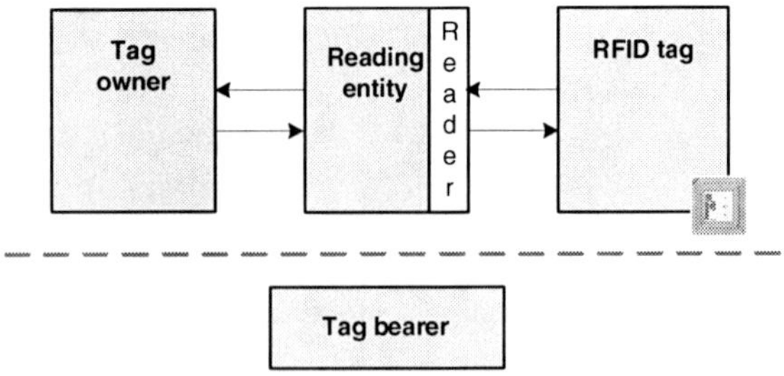

Fig. 6.4. Tag bearer not taken into account yet

But the customers who carry the tickets and that are thus the tag bearers are not respected here. This is not appropriate because they are the persons affected by unwanted recognition and tracking. In any RFID system model, either the classic one or the new one presented in the previous section, the tag owner is able to identify the tag. But in many cases, the persons affected by unwanted recognition and tracking and the tag owners that can always recognize and track the tags are not identical. This is also the case in the subway ticketing example: The company operating the subways system can decide whether data regarding tag reads is revealed or not, and the customer has no capability to influence that. Thus, the customer needs to trust the company that privacy sensitive data is processed in the way the customer would like. This necessity of trusting a company that itself has the incentive to harvest as much data as possible is not in the interest of a customer.

From a data security and privacy perspective, the tag bearer should have the capability to decide whether a tag in his possession is read or not. There are many examples in practice in which tag owner and tag bearer are not the same entity. Examples are ticketing (as shown in the example) and books that are lend from public libraries. Thus, the tag bearer that has not been taken into account yet, see the bottom of figure 6.4, should also be included in the RFID system model.

The goal should be to give the tag bearer the capability to control tag identification. If tag identification is not appropriate for privacy reasons, the tag bearer must be able to deny tag identification. This way, it is ensured that only legitimate tag reads are completed. For the tag bearer, RFID read-outs become therewith transparent which is a huge improvement compared to other RFID system models in which tags can be queried unnoticed and without approval. In other words, the tag bearer should have the capabilities of "Notice and Consent" [LDM02], i.e. he shall get informed regarding tag queries and also have the ability to choose whether to allow or to deny tag identification.

The tag bearer thus needs to act as a filter for requests for tag identification that reach the legitimate entity. Only with the consent of the tag bearer, the tag owner may be informed on a tag query and reveal data to the reading entity. The resulting system model will be explained in section 6.5. Before that, some open problems will be highlighted in the following and some looming possibilities will be explained in the next section.

Tag bearer determination

Including the tag bearer as an additional entity into the RFID system makes sense because it adheres to the practical circumstances. A problem results from getting the state in the physical world transferred to the corresponding representation in the virtual world, i.e. the RFID system. In concrete: How does the RFID system get to know who the current tag bearer is?

To adhere to the data security and privacy requirements, it is essential that the tag bearer set in the RFID system matches the actual tag bearer in the physical world. Thus, the RFID system needs to be informed about changes of the tag bearer in the physical world.

This can definitely be performed by explicit user action. But this would not be convenient and might be forgotten. Notifying the RFID system of a change of the tag bearer without explicit user action is preferred and adheres to Weiser's vision of the "invisible computer" [Wei93] much better.

The lifecycle of objects has many stages like manufacturing, distribution/relocation, stocking, use, transfer, and disposal. For a single object, a variety of different parties is involved. This also results in a number of tag bearers for a single object. Examining the whole lifecycle of some exemplary objects, one gets a clearer picture of additional requirements.

For instance, sometimes tag bearers shall remain anonymous to each other. A good example is banknotes: If banknotes are equipped with RFID tags, the central bank is the tag owner, and the people currently carrying the banknote are the tag bearers. As money in cash shall be an anonymous payment instrument, the banknotes need to be passed without revealing the identity of old and new tag bearer to each other. In this example, the tag owner must not get to know the tag bearer due to the same argumentation. This shows that tag bearers and tag owners need to stay anonymous to each other in same scenarios.

The required anonymity can be reached by decoupling entities like tag bearer and tag owners from each other using mix networks or a trusted third party. Thus, this requirement can be addressed quite straightforwardly.

The implicit change of tag bearers automatically and without the requirement for a user interaction is much more challenging but required in practical applications. Imagine a kind adult that wants to donate a lolly to a child. For such an action, nobody wants to initiate an explicit transfer of ownership. The physical transfer should suffice. Similar examples exist for the change of the tag bearer: There are simple everyday actions in daily life in which somebody moves objects on behalf of somebody else, e.g. taking a letter to a mailbox.

Ideally, a generic, user-friendly, and reliable way exists for the RFID system to notice whether a tag bearer has changed or not. For some cases in which such a change is performed in conjunction with other explicit actions, detecting a change of owner or bearer is simple, e.g. after checkout in a supermarket, the purchased items definitively have a new owner and a new bearer.

But the only generic way seems to be based on location: If an item is no longer near to a person, the person is no longer the tag bearer. If the item is near to a person, the person is probably the tag bearer or moves together with the tag bearer. In addition, the relationship of the physical location to the tag bearer can be taken into account. For example, if another person moves an item in the flat of the tag bearer, the tag bearer needs not change as long as the item does not leave the flat.

The determination of the tag bearers thus needs to be performed using context, i.e. the location of items and potential tag bearers and owners. Here, RFID research meets other research in the area of pervasive computing, e.g. context-sensing.

Besides the considerations here, the problem has been examined in a project thesis of a student, see [Kie06]. He examined objects like supermarket items, clothing, books, pharmaceuticals, tickets, and cars and considered which rights and relationships owners and bearers should have at different stages of the object lifecycle. The considerations are far from complete but have showed that the topic is sophisticated and has dependencies to social sciences. Before the design of a generic model, the actual demand needs to be well understood.

6.4 Personal Manager

As stated in the previous section, the tag bearer shall be informed of tag queries and shall be able to control whether the tag can be identified or not by the reading entity. When the RFID technology becomes ubiquitous, a large number of tag queries will be performed every day. This means that the tag bearer cannot acknowledge or deny each tag query manually. Thus, there needs to be something that performs most of such actions automatically on behalf of the tag bearer: an agent that will be called *personal manager* in the following.

The personal manger is a service acting as a user agent that can be configured by the user to act in the desired manner. The user can define policies that specify whether tag identification and the release of associated data are allowed or not. These policies need to be dependent on the context like the reading entity (identity, reader location, etc.) performing the tag query, current date/time, and current user situation. The defined user policies are enforced by the personal manager by answering requests based on the policies. In addition, the personal manager can perform logging of requests and the performed actions. Such logging information can be used by the user to adapt his privacy policies and for storing which information has been provided by the reading entity in the request for tag data. If the latter information is signed by the reading entity, the information can be used as a proof of misbehavior of the reading entity if necessary.

It must be possible to define policies in a user-friendly manner: Managing policies must be comfortable and easy to perform. Past experience with the definition of policies, e.g. firewall rules, shows that this goal is not easy to achieve and that graphical user interfaces that abstract from the low-level rules are required for convenient policy definition.

Surely, there are cases in which the personal manager is not able to decide on his own whether to release information or not. In such cases, the possibility to interact with the user would be appreciated. This can be done by using a personal device carried by the user, like a mobile phone, a PDA, or a wristwatch. Using such a device, a user can acknowledge or deny a request directly after occurrence. Note that the amount of interaction with the user should be kept as low as possible.

As it is known which user is tag bearer or tag owner of an RFID tag, each user can keep a *personal inventory*. Such an inventory is a list of items that belong to the user. Regarding each item, additional information can be kept. Each item can be regarded as an object that has a number of properties like owner, bearer, current location, description, price, or whatever else. In addition, an object history can be kept for each object.

Collecting, managing, and sharing personal information is also considered in literature. For instance, in [JHL02] repositories of personal data called "information spaces" are proposed. Physical, social, or activity-based boundaries are used there to ensure privacy-aware operation.

Personal (location) data

Besides the personal inventory, the storage and release of personal data can also be managed by the personal manager. This makes sense because context information regarding the user is required anyway to perform decisions whether to allow or to deny tag identification and the release of tag information.

The most important personal data for use in the RFID system is the user location because it helps in the automatic determination which user the current bearer of a tag is: A user that is at another location than a tag cannot be the current tag bearer.

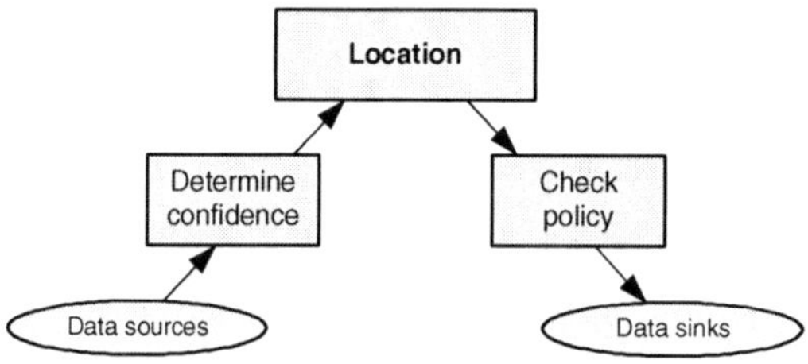

Fig. 6.5. Gathering and releasing location data

The user location can be derived by tracking an item with an RFID tag that the user carries around. A personal item like a wristwatch is suited for the purpose. If the RFID tag of that personal item is read by a reader using the presented push principle, the personal manager is being informed as the user is owner and bearer of the tag and the item. If the reader is stationary and provides information regarding its location, the personal manager regularly gets updates of the user's location this way.

Since the current user location is required, integration with other location based systems is envisioned. The personal manager is available all day so that it can also provide service for other purposes than for RFID systems only. Ideally, all means for obtaining location information, e.g. mobile phone positioning or GPS information from mobile phones and navigation systems, provide their collected location information to the personal manager. These are the "data sources" depicted in figure 6.5.

Systems or services that need the current user location can now subscribe to the required data at the personal manager. These are the "data sinks" shown in figure 6.5. As many data sources are fused at the personal manager, very precise location data can be provided by the personal manager. Of course, location data is privacy sensitive information. Therefore, it is not revealed to anybody, and the release is strictly controlled. It is also possible to release data with reduced accuracy if appropriate, e.g. the city the user resides in instead of the exact location.

Data sources should not only provide the measured location but also information regarding the precision of the data and the probability of the correctness of the measurement. This makes it easier for the personal manger to fuse the data from several data sources. Before processing the data, it should be checked for plausibility, and its confidence should be determined and respected.

Data sinks can fetch data in different ways: The data sink can retrieve the estimated current user location along with information regarding the accuracy of the data. Instead of polling, a push variant that informs a data sink on a predefined event is conceivable, too. For proximity based applications, an event can occur when the user location gets close to a predefined location. Data sinks can also ask for the probability that the user is located in a specified area.

Managing ones location data with something like a personal manager has already been proposed by other researchers in a similar manner. For instance, Myles et al. describe a system called "LocServ" that lets users automate control of their location information [MFD03]. Using user-defined policies, they want to minimize the extent to which the system needs to interact with its users.

Security considerations

The personal manager is a central representative for a user and manages personal data and the user's personal inventory. Thus, much sensitive information is stored and processed by the personal manager. Hence, security is an important aspect.

Many entities communicate with the personal manager so that much data is transmitted and processed. A critic might question whether so much data is required because from the perspective of privacy and data security; avoiding data collection is a preferable option. But for context-sensing, which is a basic building block of ambient intelligence, personal data is required. As Roßmann stated in [Roß05], the organization of communication is not intended to prevent communication but to enable it in a self-determined way that takes the interests of the individual into account.

A facility like the personal manager would of course be a good target if an attacker wanted to harm a particular user. Software bugs that cause vulnerabilities can cause serious problems for the user since the personal manager stores sensitive personal data.

From this perspective it does not seem beneficial to store a user's data in such a central manner. But on the other hand, it means that they only need to be protected at one place instead of having numerous distributed repositories in many organizations. To some extent, there is a parallel to single-sign-on: If an attacker is able to attack the system and is afterwards able to sign on on behalf of a user, he gains access to many services. On the other hand, security is increased because a user does not need to deal with many authentication systems and the probability for successful phishing attacks is decreased.

Due to the sensitivity of the data stored and processed by the personal manager, special care should be taken in its design and implementation. For instance, the personal manager should be decomposed into several independent services. These services should have their own security mechanisms instead of a single security boundary around the complete personal manager. This way, a security breach in one of the independent services cannot be abused by an attacker to access other ones.

Summary

In this section, the concept of having a personal manger that stores and processes personal data and that keeps a personal inventory has been presented. An overview is shown in figure 6.6.

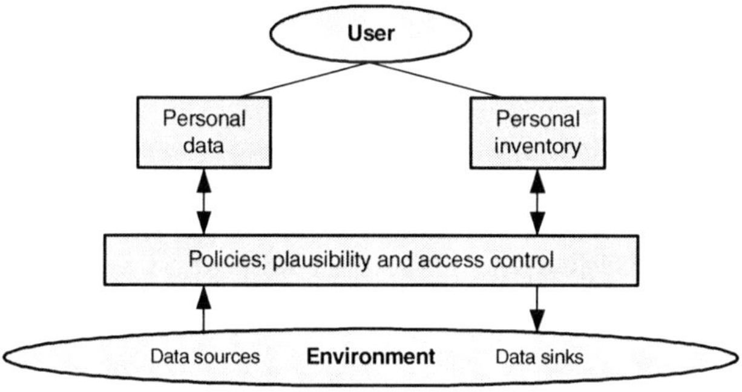

Fig. 6.6. The personal manager controlling personal data and keeping a personal inventory

One service of the personal manager is controlling the collection and release of personal data like the user's current location. Another service is keeping a personal inventory that lists all the objects with RFID tags owned or carried by the user. Communication with the environment, i.e. data sources and data sinks, is performed using defined interfaces and is controlled via policies and checks.

6.5 Assembling the Building Blocks

Within this chapter, the classic RFID model has been modified and extended. In this section, the new model, the RFID protocols from chapter 4, and the pseudonymization infrastructure introduced in chapter 5 are assembled to implement a complete RFID system. Figure 6.7 shows an overview of the system with all involved entities: RFID tag, reading entity with RFID reader, a pseudonymization infrastructure, current tag bearer, and current tag owner.

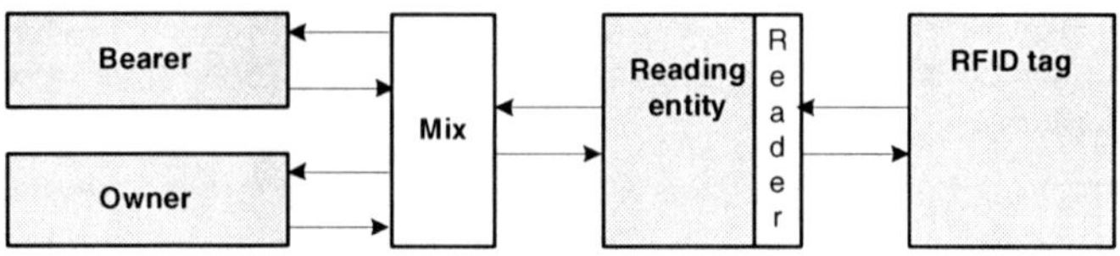

Fig. 6.7. Overview of the entities in the extended RFID model

The procedure in the assembled system is as follows: The RFID tag is queried by a reader so that the respective reading entity obtains the tag's current backend identifier and tag identifier. These identifiers do not contain any information usable by the reading entity. Using the pseudonymization infrastructure, the reading entity thus needs to forward the identifiers and a request for the tag's data to the tag bearer. Based on the received information, the tag bearer decides whether the request is legitimate or not. In the next step, the tag owner can decide whether the request is legitimate or not. If both, tag bearer and tag owner, agree to the request, the reading entity obtains the requested data regarding the tag. In any case, the reading entity gets a reply message that includes the message that is forwarded to the tag and initiates the identifier modification there.

Tag bearer and tag owner can process the data regarding the tag query. They can also forward the data to other organizations like subcontractors. The data is managed using personal managers, one for the tag bearer and one for the tag owner. The personal managers check privacy and information release policies and perform the data sharing with other organizations.

Evaluation

For the evaluation of the RFID system, the stated process of tag read-outs is assumed to make use of the RFID protocol implementation presented in chapter 4 and a pseudonymization infrastructure as presented in the previous chapter. The overall system characteristics thus results from the properties of these two components.

Regarding security and privacy, the resulting system has very good properties. The reason is that both, the RFID protocol as well as the pseudonymization infrastructures, have been designed with security and privacy as main goal. A small drawback compared to ideal properties is that the tag requires the reception of the reply message to modify its tag identifier. This can be mitigated by using the pseudonymization infrastructure not only for backend but also for tag identification. The privacy properties of the pseudonymization infrastructure depend on the design decisions taken in the implementation. The layered approach of the pseudonymization infrastructures presented in the previous chapter implements the principle of shared trust. The tag bearer is part of the RFID system and can decide for each tag read-out whether tag data is revealed or not. This makes system operation transparent to the tag bearer and gives him the opportunity to protect his privacy effectively.

The resource consumption of the two components has already been discussed separately for each component. The resulting resource requirements can be approximated by the sum of those of the individual components. As stated in subsection 4.8.5 in which the "Hash-based ID variation" scheme has been evaluated, that scheme has good characteristics regarding resource consumption. Regarding the pseudonymization infrastructure, the characteristics are not that good. Even if an infrastructure based on hash functions as cryptographic primitive is employed, generating a pseudonym requires the calculation of multiple hash values, one for each level of the infrastructure topology, in the tag in each tag query.

Scalability has been the reason for using pseudonymization infrastructures in RFID systems. The result is that the system has good scalability properties. The only deficiency is the root node of the pseudonymization infrastructure that can become the bottleneck of the system.

In both system components, different hash functions can be employed in parallel. This way, a migration path to newer algorithms is given. The RFID protocol does not store shared keys so that no attack targets that could affect more than a single tag are given. For this aspect, sustainability is very good. The long-term characteristics of pseudonymization infrastructures depend on the implementation. As discussed in the previous chapter, hash-based pseudonymization infrastructures in the current state of research cannot be expected to provide privacy protection in the long term.

As already stated, much computation is required in the RFID tags. On each tag read, three hash calculations need to be performed for the RFID protocol and one hash calculation per level of the pseudonymization infrastructure. Although a precalculation is possible, this is not a good result for resource scarce tags that need to provide query results within a short timeframe. Both components, i.e. the RFID protocol and the pseudonymization infrastructure, have the disadvantage that no delegation or caching is possible. This means that on each tag read all entities need to be involved. An advantage of the system is that only a single message exchange is required for all tasks, i.e. tag identification, mutual authentication, and identifier modification. But the roundtrip time within this message exchange is large because the messages need to pass several entities and also all the levels of the pseudonymization infrastructure.

Regarding handling and practicability, the system has serious drawbacks. Due to the many calculations that are required, tags cannot be read out fast. Therewith, users would have to wait in certain actions. This does not lead to a good user experience. Another problem in practice would surely be reliability. The pseudonymization infrastructure needs to be operated reliably, but there are no direct incentives for organizations to do so. An advantage of the system as presented is that no alternative channels to the wireless communication with the RFID tag are employed so that no explicit user action is required.

The system with the presented building blocks is application independent. It just provides a secure, privacy respecting infrastructure for tag identification and authentication. Application specific demands can be addressed on top of this infrastructure flexibly. Therewith, the system is universally employable.

In contrast to the RFID protocols presented in chapter 4, a system with the additional entities presented in this chapter has a much larger scope: It can operate inter-organizationally. This fits the practical requirements.

Conclusion

The evaluation showed that an RFID system that involves all presented entities and that is technically based on an RFID protocol as presented in chapter 4 and a pseudonymization infrastructure as presented in chapter 5 matches the security requirements and respects the interests regarding privacy of the involved entities. There are no proposals in the literature that provide such characteristics in inter-organizational RFID systems in such a scalable manner that intends use on a global scope. Considering the tag bearer as separate entity in the RFID system is highly innovative as well.

Regarding economics and practicability, the presented solution does not satisfy the demand. The resource consumption and the requirements regarding performance exceed the capabilities of low-cost RFID tags. In addition, the presented pseudonymization infrastructures are not mature enough for practical application: Regarding reliability and sustainability much more research would be required before.

Nevertheless, the presented solution shows that building inter-organizational RFID systems that can operate in a global scope and that adhere to the security and privacy requirements are possible in theory. It also shows that only the implementation of a hash function is required in tags to reach the goals regarding security and privacy. In the next chapter, more practical solutions will be presented.

6.6 Summary

In this chapter, the RFID system model has been extended to support inter-organizational RFID systems that adhere to the identified security and privacy requirements.

At first, the classic RFID model has been presented that includes legitimate and illegitimate reading entities as well as the RFID tags. RFID protocols like the ones presented in chapter 4 aim at securing the communication between legitimate entities and the tags. But the model does not address the requirements that arise in inter-organizational RFID systems.

The first modification of the classic model introduces reading entities that are regarded untrusted at first. Whether information regarding a queried tag is revealed to such a reading entity or not is decided by the tag owner, i.e. the entity in charge of the tag. This proceeding leads to a push concept: Each reader, whether legitimate or not, needs to inform the tag owner regarding a tag query to obtain usable information. This way, readers become shared infrastructure components and obtain data for arbitrary tag owners from arbitrary tags. Data sharing between organizations becomes much easier this way because data can be shared on RFID system level instead of having to use application-specific interfaces.

The second modification of the classic model has been the introduction of the tag bearer as additional entity. The tag bearer is not involved in the classic model but he is the one who is concerned regarding the violation of location privacy and the release of personal data. Thus, it makes sense to consider him in the system, too. The main problem is the automatic determination which user the current tag bearer of a tag is. This refers to the research area in ubiquitous computing that deals with context-sensing.

Tag bearer and tag owner need to manage their tags. An agent called *personal manager* that acts as an agent on behalf of the user has been proposed. The personal manger manages personal data like the current user location and a personal inventory. Data is released to extern entities based on user-defined policies. In exceptional cases in which a special user acknowledgement is required, the personal manager can interact with the user using a personal device that the user carries.

Finally, the new RFID model, the RFID protocols from chapter 4, and the pseudonymization infrastructures from chapter 5 have been assembled to a complete RFID system. An evaluation showed that the system meets security and privacy requirements well but that it is not mature enough for practical application. This leads to the challenge to develop more practical solutions that better adhere to the economic requirements. Addressing this challenge is the goal of the next chapter.

7

Current Research

The complete RFID system presented in the previous chapter has very good characteristics regarding the protection of security and privacy, but its applicability in practice is not given. In this chapter, advanced solutions are developed that do not only focus on security and privacy but also more on practicability and economic efficiency.

Partial solutions are presented in the next section. At first, a more elegant alternative to the "Hash-based ID variation" approach that has been shown in section 4.8 is introduced. Second, a scheme for securing supply chains against product counterfeiting is developed. This scheme does not provide location privacy. But in return, it does not require the implementation of a cryptographic primitive in RFID tags so that the scheme is very inexpensive and thus economically very interesting.

In the subsequent section, a new overall RFID system architecture is presented. It is an alternative to the one shown and evaluated in section 6.5 at the end of the previous chapter. In contrast to the solution assembled there, the new *ID-Zone Architecture* better matches the practical privacy requirements derived in chapter 2. Hence, the new architecture has better characteristics and is more practical and economic.

7.1 Partial Solutions

In this section, solutions are developed that only address the identified issues in part. In return, the solutions better adhere to practical requirements. In the following subsection, a more elegant alternative to the "Hash-based ID variation" approach is presented. In a second subsection, a scheme for securing supply chains against product counterfeiting is developed.

7.1.1 Identifier Modification Based on Triggered Hash Chains

Ohkubo et al. proposed to modify tag identifiers using hash chains [OSK03]. The scheme has already been derived and discussed in subsection 4.5.3. It belongs to self-contained modification based on dynamic data since it does not require communication for identifier modification and does not rely on lists of static data.

In the hash chain approach, each tag has an inner state that is never revealed to the outside. This inner state is updated in each modification operation using a hash function: The current inner state is the input of the hash function; the new inner state is the hash value. From the current inner state, the current tag identifier that is revealed to the outside world is calculated using a second hash function.

The scheme provides indistinguishability and forward secrecy. From a security and privacy perspective, it has very good characteristics. The main problem of the scheme is to keep tag and responsible backend entity in sync: Lost messages or tag queries by an attacker bring the two peers out-of-sync. The backend entity needs to perform many iterations of hashing to all stored inner states of tags until the tag can be identified.

If the number of iterations on the backend side is limited by policy, an attacker can render a tag unidentifiable by enough repeated queries. If no limit is applied, an attacker can run a denial-of-service attack against the backend entity by inserting an invalid tag identifier into the system. A solution is to introduce a limit of iterations that a tag performs before it starts over with its initial identifier. But then identifiers repeat and the property of indistinguishability is no longer given. This could be abused by an attacker for unwanted recognition and tracking.

In any case, the approach is not well scalable since many hash operations are required to identify a single tag. The complexity is $O(n^2)$, whereby n is the number of tags known to the backend entity. Optimizations discussed in the literature like Avoine's "time-memory trade-off" [AO05b] mitigate the problem to some extent but do not solve it since it is inherent. Thus, Ohkubo's approach cannot be used in practice, but it is a good conceptual base for creating other protocols.

The *Hash-based ID variation* approach has been discussed in section 4.8. This scheme is based on message exchanges and does not have the synchronization problem between tags and responsible backend entities. This is accomplished by adding Δtid in the first message (see figure 4.24). A Δtid value different from one clearly indicates a failed identifier modification. This value can be used by the entity responsible for the tag to detect communication failures and attacks, but on the other hand, it indicates to an attacker that no identifier modification has been performed. The scheme is rather simple and clear compared to other schemes, but it nevertheless has a complexity that provides pitfalls for performing secure implementations, see section 4.8.3.

In the following, the hash-based ID variation approach is combined with the hash chain concept. The result is called *Triggered hash chain* approach. The goal is to have a scheme that has the same desirable properties as hash-based ID variation, but that lacks the inelegant Δtid and the resulting issues. In parallel to the publication of this book, the protocol is presented and published at the Sixth Annual IEEE International Conference on Pervasive Computing and Communications, PerCom 2008, see [HM08].

Description of the triggered hash chain approach

The basic idea is to perform an update of the inner tag state using a function like in self-contained modification based on dynamic data. In contrast to the pure self-contained approach, an update is not performed on each tag query but only when triggered by the responsible backend entity. Therewith, the scheme presented here belongs to the approaches based on a message exchange.

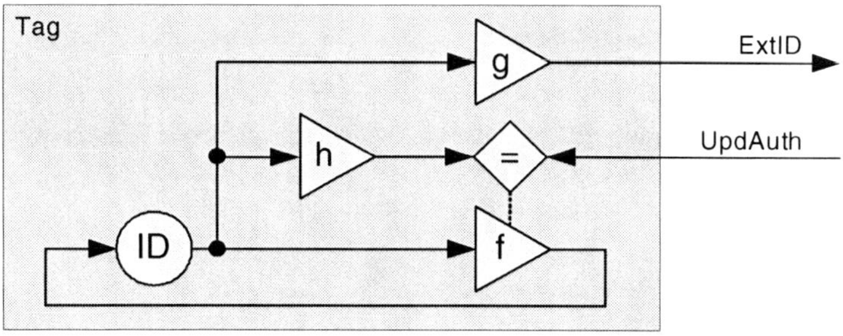

Fig. 7.1. Tag organization in the triggered hash chain approach

Figure 7.1 depicts the tag organization for the new approach. The tag has an inner state *ID* that is never revealed to the outside world. The responsible backend entity controls the tag and knows that inner state.

The current tag identifier is calculated by applying the hash function g, i.e. by calculating $extid = g(id)$. As the responsible backend entity knows the inner state of the tag and can also calculate $g(id)$, the backend entity can identify the tag successfully.

The responsible backend entity now attempts to trigger an update of the inner tag state. Therefore, the backend entity calculates $updauth = h(id)$ using the hash function h and sends the result via the reading entity back to the tag. The tag also calculates $h(id)$. If the result and the received value match, the inner state of the tag is updated by calculating $id^* = f(id)$. No modification of the inner state takes place in the other case, and the received message is discarded by the tag. The function f needs not have the one-way property of a hash function, but a hash function provides forward secrecy.

Analogously to the hash-based ID variation approach, the responsible backend entity always keeps two records for each tag in its tag database. If an update message gets lost, the backend entity can therewith still identify the tag using the old external tag identifier *extid*. Like in the hash-based ID variation approach, both records point to each other and the respective unused record is overwritten by an update.

A variant of the approach is shown in figure 7.2. Instead of having two separate hash functions g and h, only one hash function h is employed in a slightly different setup. The hash function is applied twice to calculate the external tag identifier, i.e. $extid = h(h(id))$. In the next step, the responsible backend entity returns $updauth =$

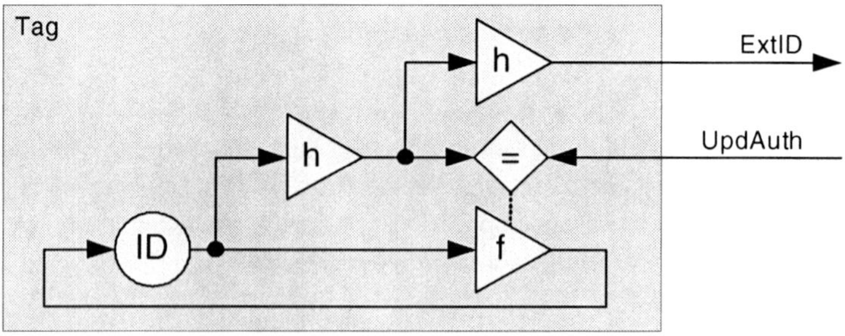

Fig. 7.2. Variant of tag organization in the triggered hash chain approach

h(*id*) to trigger the update of the inner tag state. The possibility to calculate *extid* when *updauth* is given is no insecurity because the external tag identifier *extid* has been revealed to the outside world before anyway.

Protocol cycle and protocol analysis

In this paragraph, an example of the protocol operation is presented. Afterwards, the susceptibility of the protocol regarding message loss, message interception, and replay attacks is analyzed. The different cases are illustrated by means of examples.

Tag	*Legitimate entity*		
ID	*g*(*ID*)	*ID*	*PTR*
id	*g*(*oldid*)	*oldid*	*g*(*id*)
	g(*id*)	*id*	*g*(*oldid*)

Table 7.1. Valid state between tag and legitimate entity as starting basis

In table 7.1, a usual synced state between tag and its responsible backend entity is depicted. The tag only has a single state variable, i.e. the internal tag identifier *ID*. This internal identifier has a value of *id* in the example. The legitimate entity has a database with a table for keeping data regarding its tags. The table has hash values *g*(*ID*) as table keys and also has a field for the internal tag identifier *ID* and a *PTR* field for pointing to related records. This is similar to the organization in the hash-based ID variation approach, but the transaction identifier fields are not required here. The tag table contains two records for each tag. One record belongs to the current tag state. This is the record with *g*(*id*) as table key, *id* as internal tag identifier and the pointer to the second record. That second record contains the data belonging to the previous tag state. This data is kept for the case of message loss, just like in the hash-based ID variation approach. The old internal tag identifier is depicted abstractly as *oldid* because the concrete value is not relevant for the example.

REGULAR TAG QUERY: In a successful protocol run, the tag is queried and yields *extid* = *g*(*id*) as external tag identifier. The legitimate entity can identify the tag by

Tag	Legitimate entity		
ID	$g(ID)$	*ID*	*PTR*
$f(id)$	$g(f(id))$	$f(id)$	$g(id)$
	$g(id)$	id	$g(f(id)$

Table 7.2. State after regular update

the corresponding record entry in the database. If not already done, the legitimate entity overwrites the second record with $f(id)$ as internal tag identifier and $g(f(id))$ as external tag identifier or table key, respectively. Afterwards, the legitimate entity sends $updauth = h(id)$ back to the tag. The tag compares the received *updauth* with its calculated $h(id)$ and updates the internal tag identifier by calculating $id^* = f(id)$ if both values match. The resulting state in tag and database is depicted in table 7.2.

FIRST MESSAGE LOST OR INTERCEPTED: The first message of a protocol cycle transports the current external tag identifier from the tag to the responsible entity. This message can get lost due to an error, or an attacker can prevent the message from being delivered. The message loss itself does not cause any harm because neither tag nor legitimate backend entity change their state due to the tag query. Three cases can be distinguished for the whereabouts of the first message:

- Message is never inserted again.
- Attacker inserts message again before a successful protocol cycle between tag and legitimate entity occurred.
- Attacker inserts message again after a successful protocol cycle between tag and legitimate entity occurred.

The first case, in which the message with the external tag identifier is never inserted again, is uninteresting because the tag state did not change due to sending the message. With respect to state information, loss of the message is thus transparent for tag as well as backend entity.

The second case, in which the message is inserted again before an answer from the backend entity reaches the tag, leads to the creation of messages with the same content in successive tag queries. This can lead to several messages containing the same external tag identifier reaching the backend entity. Upon the first reception of such a message, the legitimate entity updates the record with the tag identifier that has not been used for identifying the tag. This is the intended behavior within the normal protocol operation. When a message with the same content reaches the backend entity again, no change of the backend state occurs because the same update would be performed again and would not change anything.

The third case is the most interesting one because it has itself three cases that need to be distinguished. The completion of the protocol cycle causes the tag state to be updated so that a new external tag identifier is emitted by the tag in the future. If the old, intercepted message reaches the backend before a message with such a new external tag identifier has reached the backend, there is no difference to the previously described case: The old message does not cause a state change at the backend entity. In the second case, in which the old message reaches the backend after the new external identifier has reached the backend, again no state change in the backend occurs. The backend will only reply with a message triggering an update of the tag state that has already been performed so that that reply message gets discarded by the tag. A third case is the insertion of the old message after two protocol cycles have been completed successfully and the new external tag identifier has reached the backend. As the backend database always contains a record with the current and one with the previous external tag identifier, the external tag identifier contained in the old message is no longer known to the backend entity. Hence, the old message is discarded by the backend entity and no other action is performed.

SECOND MESSAGE LOST OR INTERCEPTED: The second message is the one sent from the backend entity back to the tag to trigger the identifier modification. This message can get lost either by an error or by malicious attacker activity.

Tag	*Legitimate entity*		
ID	$g(ID)$	ID	PTR
id	$g(f(id))$	$f(id)$	$g(id)$
	$g(id)$	id	$g(f(id)$

Table 7.3. State after loss of second message

Table 7.3 shows the state in tag and backend after loss of the second message. As one can see, one record in the backend was updated upon reception of the first message. But as the second message got lost, no update of the internal tag identifier took place.

Again, the already known three cases can be distinguished for the whereabouts of the second message:

- Message is never inserted again.
- Attacker inserts message again before a successful protocol cycle between tag and legitimate entity occurred.
- Attacker inserts message again after a successful protocol cycle between tag and legitimate entity occurred.

If the message is never inserted again, the tag keeps its current internal tag identifier. Hence, the tag will provide the same external tag identifier again in the next tag query. The corresponding message will reach the backend entity and the tag will be identified using that old external tag identifier. As the update of the second record has already taken place, the backend state will not change any more caused by the reception of an identical message. The backend will create a second message identical to the lost one. This can be regarded as a retransmission. For this retransmitted second message, the same three cases as described here hold.

The second case is the insertion of the message before another protocol cycle between tag and legitimate entity is completed. The inserted old message will trigger the intended update of the internal tag identifier. Compared to the regular protocol operation, the later insertion of the second message therewith results just in a delay.

If in the meantime the tag has been queried again, the backend entity will send an identical message as reply to the tag. This message will be discarded by the tag since the identifier modification triggered by the included hash value has then already been performed triggered by a previous message.

If the message is delivered to the tag after successful completion of another protocol cycle, the message is discarded by the tag. The reason is that a duplicate of that message has reached the tag earlier so that the identifier modification triggered by that message has already taken place. The message will thus not match the one expected to trigger another update and is discarded.

SUMMARY: The previous discussion illustrated the operation of the triggered hash chain protocol. At first, the normal operation of the protocol has been shown by means of an example. Afterwards, error and attack scenarios have been discussed: message loss, message interception, and message replay. The considerations showed that the protocol is able to deal with such circumstances. No such event can bring tag and legitimate backend entity out-of-sync. Thus, the protocol always operates reliably.

Evaluation

In this section, the triggered hash chain protocol is evaluated according to the evaluation criteria presented in section 4.7. As the characteristics of the triggered hash chain protocol are very similar to the hash-based ID variation approach (see the evaluation in subsection 4.8.5), the discussion here is kept relatively short and only the differences to that approach and the most important aspects are highlighted.

SECURITY: The triggered hash chain protocol meats the security goals similarly to the hash-based ID variation protocol. Data security is maintained by keeping data associated to the tag in the backend. Counterfeiting is prevented by keeping an inner tag state so that the tag cannot be cloned. A secure tag authentication that excludes the possibility of a replay attack can be performed by querying the tag twice: The first read identifies the tag. The second read is used to ensure that a complete protocol cycle has been performed: The tag reveals a new external tag identifier in the second read. As it is based on the inner tag state, the ability to provide this new identifier authenticates the tag. Illegitimate access to the system can be prevented in the same way by only treating identification data as valid after secure tag authentication. Unwanted recognition and tracking is prevented by identifier modification. Only the legitimate backend entity has the required information to identify a tag by its external tag identifier after a performed identifier change. Attacks that lead to a denial-of-service are prevented by protocol design: Attacks like message interception or replay attacks cannot bring tag and responsible backend entity out-of-sync to thus disturb proper system operation.

Regarding additional properties, the triggered hash chain protocol is advantageous. Forward secrecy can easily be provided by implementing the function f using a hash function. Further, the only inner tag state is the internal tag identifier. This lessens complexity compared to the hash-based ID variation protocol so that the triggered hash chain protocol can be expected to be less vulnerable to attacks and to be easier to implement. As a state change in the tag can only be triggered by an appropriate message that just the backend entity can craft, an attacker cannot enforce a state change for using such a state transition for attacks. In contrast, each tag query causes a tag to change its state in the hash-based ID variation protocol so that an attacker gets a lot of data and has more possibilities for attacks. Just like the hash-based ID variation protocol, the triggered hash chain protocol does not share secrets among several tags. Hence, the incentive for an attacker to perform advanced attacks like physical extraction of secrets is low.

	HIDV	HIDV variant	Hash chain	THC	THC variant
Storage					
Tag	$3l$ ($5l$)	$3l$ ($5l$)	$1l$ ($2l$)	$1l$ ($3l$)	$1l$ ($3l$)
Reader	none	none	none	none	none
Backend	$2 \cdot 5l$	$2 \cdot 5l$	very much	$2 \cdot 3l$	$2 \cdot 3l$
Computation					
Tag	$3h$	$4h$ ($3h$)	$2h$	$3h$	$3h$
Reader	none	none	none	none	none
Backend	$3h$	$4h$ ($3h$)	very much	$3h$	$3h$
Communication					
Tag → Backend	$3l$	$3l$	$1l$	$1l$	$1l$
Tag ← Backend	$2l$	$1l$	$1l$	$1l$	$1l$
Features					
Precalculation	everywhere	everywhere	everywhere	everywhere	everywhere
Forward secrecy	largely	largely	full	full	full

Table 7.4. Comparison of different approaches

RESOURCES: A comparison regarding memory use, required computation, and amount of data exchanged in communications is shown in table 7.4 in the style of the protocol comparisons shown in [LHLL05]. *HIDV* denotes the hash-based ID variation protocol, *THC* the new triggered hash chain protocol. The size l denotes the size of a single variable in the comparison of memory use and the comparison of the amount of data. The computational effort h denotes the effort for a hash operation. Effort for other operations can be neglected. The values in brackets at the tag storage indicate the memory requirement if precalulation (see below in the performance paragraph for more information) is performed.

As can be seen, the triggered hash chain approach outperforms the hash-based ID variation approach in all disciplines. Albeit uninteresting for practical application, the pure hash chain concept is also shown in the table as a reference. It requires one hash operation less than the new triggered hash chain protocol in the tags but is impractical in the backend.

SUSTAINABILITY: With a slight extension, the triggered hash chain protocol can use different hash functions simultaneously: A tag is identified by the responsible backend entity by the external tag identifier, i.e. a hash value calculated using an arbitrary hash function. The backend entity needs to have an additional field in its database that denotes which hash function to use for the respective tag. This way, the required hash operations can be performed with the hash function that matches the one implemented in the tag. The ability to use different hash functions in parallel provides a migration path to newer, more secure hash algorithms.

A broken cryptographic primitive gives an attacker control that he would normally not have. In the worst case, the attacker could bring tag and backend out-of-sync and could get the ability to clone tags. Due to the lower complexity of the triggered hash chain protocol compared to the hash-based ID variation protocol, there is probably less target area for attacks. As there are no shared secrets amongst tags, the impact of a revealed inner tag state is restricted to a single tag.

PERFORMANCE: As shown in table 7.4, in the triggered hash chain protocol, resource consumption is lower than in the other protocols. This leads to very good performance: Tags can be queried and their tag identifiers be modified comparatively fast because not much data needs to be transmitted and only few hash calculations need to be performed at each peer.

As shown in the table, precalculation is possible in all the protocols. In practice, one will thus probably spend additional memory for caching. This means that hash operations can be performed before requiring the result. The result of these hash operations is then cached until it is needed and can then be readily delivered without waiting for calculations to be completed. This is advantageous regarding performance.

HANDLING/PRACTICABILITY: The operation of the hash-based ID variation protocol and the triggered hash chain protocol is quite similar. For instance, no alternative channels which would require user interaction are used in both protocols. But the better performance of the triggered hash chain protocol improves handling because operations are finished in a shorter time.

In the hash-based ID variation protocol, tags need to stay powered between the tag query and the delivery of the reply message originating from the legitimate backend entity. Otherwise, the reply would be discarded by the tag due to a non-matching transaction number. This is different in the triggered hash chain protocol: A tag can be queried and the communication with the responsible backend entity can be performed in the background without the tag being online. The reply message can be cached by the reading entity and be delivered directly after the next query of the tag. This way, a high speed of reading, which is for instance required at band-conveyors, can be reached. This makes the triggered hash chain protocol much more practical than other protocols.

SCALABILITY, UNIVERSALITY, AND SCOPE: Regarding these criteria, the triggered hash chain approach is very similar to the hash-based ID variation approach. The information provided in subsection 4.8.5 is therewith also adequate here and is thus not repeated.

Summary

In this subsection, an approach called *Triggered hash chains* has been presented. It is an alternative to the *Hash-based ID variation* presented in chapter 4. Both have the same scope of application. But compared to the hash-based ID variation approach, the protocol presented here is more elegant: It is less complex, requires less memory in tag and backend, and has smaller messages. It also allows tags to get offline while the reading entity communicates with the responsible backend entity.

7.1.2 Policy Restricted Key-Value Pair Authentication

As stated in the last example in section 3.1, product counterfeiting is a huge problem. RFID tags that cannot be cloned easily and that are able to perform tag authentication can mitigate this problem. Such tags could be applied to products at risk. If product and tag are inseparable, authentication of the tag also proves the genuineness of the product. This subsection will present a lightweight approach for tag authentication. Location privacy cannot be ensured with this approach, but in return, the approach does not require the implementation of a cryptographic primitive in tags.

Towards key-value pair authentication

In section 4.5.2, approaches for single message authentication based on static data have been presented. The general idea has been that tags have a list of authentication values that are only known to the respective tag and the responsible backend entity. The domain of these authentication values needs to be that large that guessing an authentication value can only be performed with negligible probability. When a tag is capable of presenting one of these authentication values, one can be quite sure that the tag is genuine.

As an extension, one can make the authentication values be indexed. Therewith, the backend entity can request to reveal a specific authentication value instead of an arbitrary one out of the list. This reduces the probability of a correct guess further. With the index as the challenge and the corresponding authentication value as response, this is a simple challenge-response protocol so that the protocol belongs to the class of authentication protocols based on message exchange.

It needs to be ensured that only the legitimate backend entity can bring the tag to revealing an authentication value. Therefore, the backend entity needs to prove its authenticity so that a mutual authentication scheme is required ultimately. Having a single key that makes the tag reveal an authentication value is not a good solution because the key could be overheard by an attacker upon transmission and be replayed later for thus revealing the other authentication values. A possible solution is not to use a simple sequential index, e.g. starting from zero and counting up. Instead, randomized index values can be used as keys of the authentication values. Just like the authentication values, these keys need to have a domain that is large enough so that the probability of a correct guess by an attacker is negligible.

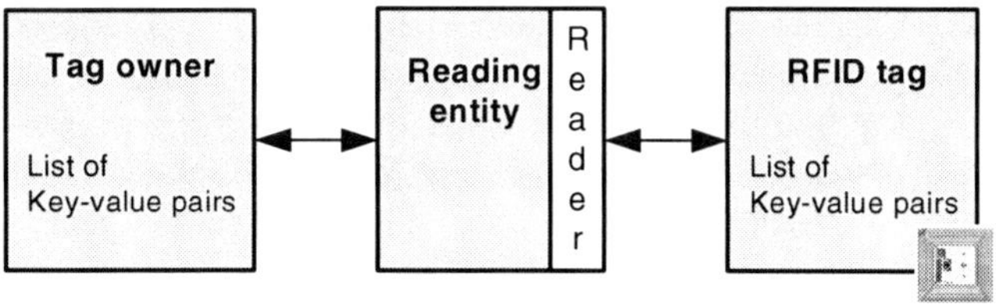

Fig. 7.3. Tag authentication with key-value pairs

Thus, each RFID tag and its respective tag owner have a list of number pairs, see figure 7.3. Each pair consists of a key and an authentication value. The third involved entity is the reading entity that wants to check whether the tag is genuine or not. As stated in the previous chapter, this entity should be regarded untrusted because it might be an illegitimate reading entity or even an attacker. Thus, the reading entity may not get to know the list of key-value pairs.

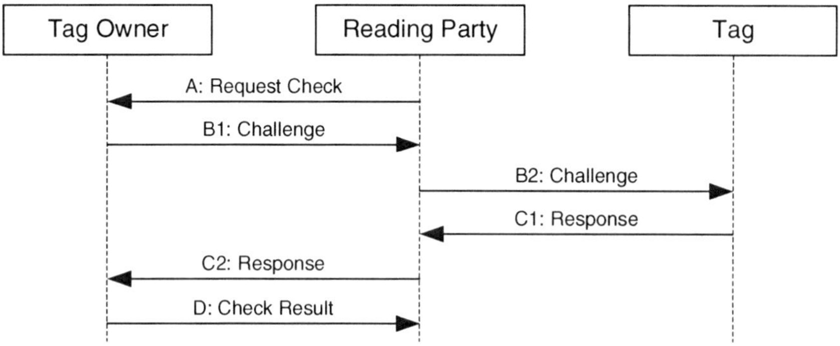

Fig. 7.4. Procedure of tag authentication with key-value pairs

Figure 7.4 depicts the procedure of an authentication process. It is assumed that tag identification has already taken place so that the tag owner knows for which tag the authentication shall be performed. When the reading party wants to check the genuineness of the tag, it sends a first message *A* to the tag owner requesting the check. The tag owner selects a key-value pair from its list and sends the key via the reading party to the tag. This is the challenge and shown as messages *B1* and *B2* in the figure. If the received key matches a key in the list, the tag returns the corresponding authentication value via the reading party to the tag owner, see messages *C1* and *C2*. If the received authentication value matches the one stored in the tag owner's list, the tag owner regards the tag as genuine and returns a positive check result via message *D* to the reading party.

The tag owner needs to ensure that the received response in message *C2* corresponds to the challenge sent in message *B1*. An authentication value that belongs to another key-value pair needs to be rejected. The tag owner should mark key-value pairs whose key has been sent in a challenge message as used. The reason is that after revealing a key, an attacker can obtain the corresponding authentication value by querying the tag and becomes therewith able to mimic the tag in such a way that the mimicking device or a tag clone passes an authentication process in which that key is used. Used key-value pairs should never be reused for that reason. Each key-value pair is thus only suited for a one-time authentication.

This is the deficiency of an authentication scheme that relies on static data. As already explained in subsection 4.5.2, the key material becomes depleted sometime so that no reliable authentication is possible any more. Correspondingly, in the approach presented here, the number of key-value pairs stored in tag and backend entity is limited so that only a fixed number of authentication processes can be performed. After all key-value pairs have been revealed, a perfect clone of the respective tag can be created so that it is no longer possible to differentiate between a genuine tag and a cloned one.

A straightforward measure against depletion of authentication pairs would be to store a high number of such pairs on each tag. But this measure is not a good one for two reasons: First, non-volatile memory is required on the tags for each key-value pair. Hence, more memory on the tags is required for a higher number of pairs. This makes the tags more expensive. Second, a higher number of pairs mitigates the problem but does not solve it. An attacker can still perform the required number of plausible requests for tag authentication to reveal all the key-value pairs of a tag. Requiring the requestors to reveal and prove their identities to counteract this attack does not work because customers should be able to check the genuineness of products anonymously. An attacker could use such anonymity to obtain all the key-value pairs of a tag and ultimately clone that tag.

Policy-based control

There seems to be no generic solution to the stated issue. But for particular application areas, one can make use of given application area characteristics. Product counterfeiting takes place along the supply chain. Thus, one can take a closer look at the procedures along supply chains and the parties that are involved there. The identified characteristics can then be used for creating a solution specific to supply chains.

The route of products along the supply chain can be very different. It is unexpectedly long in some cases, e.g. in the case of international chains crossing several countries. Different prices of some products in different countries even lead to repackaging so that the products intended to be sold in one country can be sold profitably in countries where the product price is higher. There are also markets where surplus stocking is sold. Hence, one cannot reduce supply chains to a number of usual paths because there is more complexity in reality, even in pharmaceutical supply chains or supply chains of other highly regulated business sectors.

But albeit the unmanageable number of different paths that a product may take between manufacturer and consumer, classes of involved parties are quite manageable. All relevant involved parties along the supply chain can be assigned a certain role. For instance, one can distinguish manufacturers, distributors, wholesalers, retailers, customs, commercial or industrial final consumers, and individual final consumers.

The core idea of policy-based control is to reserve particular key-value pairs for parties fulfilling a certain role. For instance, two pairs could be designated for exclusive use by customs. If a product reaches a customs office, the latter could request the tag owner to perform a check for authenticity of the product. In the request message A, see figure 7.4, the customs office would need to prove its affiliation to the customs role. The tag owner would then use one of the pairs reserved for use by customs to perform the check for genuineness and mark the key-value pair as used. In practice, the tag owner would also log the identity of the requestor along with a timestamp.

The tag owner thus controls and limits the use of key-value pairs based on the role performed by the requesting reading party. When the pairs that have been reserved for a certain role are depleted, the tag owner will not allow another check for genuineness using another fresh, yet unused pair. This prevents attackers from learning all pairs and therewith from becoming capable to clone the respective tag.

In case that a check for genuineness cannot be performed with a fresh key-value pair, a key-value pair can be reused. Such a check with a used pair does not provide full security so that the reading party should get a notice that even a positive check result is only an indication but does not reliably ensure the validity of the product.

In repeated requests for checks of the same tag by the same reading entity, the tag owner should always use the same key-value pair for the check. This way, the available pairs do not get depleted too fast. From a security perspective, the reuse of pairs by the same reading entity is tolerable because eavesdropping by extern attackers is unlikely due to the physical security provided by business premises.

In some cases, key-value pairs that have been reserved for parties acting in a certain role can be used for authenticity checks by parties acting in other roles. For instance, after a check has been performed by an end-user, one can assume in many branches of trade that the product will never return to a distributor.

Some of the key-value pairs of a tag should be reserved for anonymous checks for authenticity. Preventing depletion of fresh pairs in the scenario of anonymous checks is more difficult because there is less control and nobody to be held liable for the actions. Thus, it is a challenge to find a working policy for such an anonymity scenario.

One approach would be a rigorous limiting of fresh key-value pairs. For instance, one could only allow a single check for authenticity with a fresh pair. In successive anonymous requests, old pairs would be reused. Another approach would consider at which party the last check for authenticity has taken place: If the last check was an anonymous one, a pair would be reused for the current check. If the last check was one by a retailer, one would expect that the current check would be performed by an end-user and allow this check using a fresh pair. A third approach is based on alternative channels, see section 4.6.3. A code can be printed on the packaging of the product. Anonymous requests for checking genuineness can now be required to be authorized by this code. If no other similar request was received within a reasonable span of time, the tag owner would then perform a check for genuineness using a fresh key-value pair. In this scenario, the code that can only be gathered optically prevents wireless attacks effectively.

Physical uncloneable functions

As presented in section 4.5.2, instead of keeping a list of key-value pairs in memory, a cryptographic primitive like hash functions can be used to calculate authentication material when needed. But the requirement to implement a cryptographic primitive was avoided in the presented approach to keep RFID tags inexpensive.

So-called *physical uncloneable functions* (PUFs) might become an interesting alternative to storing lists of key-value pairs in tags. Such a function is dependent on a physical device, maps challenges to responses, is easy to evaluate, and is hard to characterize [GCvDD02]. The dependency between such a function's preimage (challenge) and its output (response) depends on physical characteristics of the device.

Silicon PUFs (SPUFs), see [GCvDD02] and [LLG⁺04], are suited for use in RFID tags. The idea of this kind of PUFs is to make use of unique delay characteristics of each integrated circuit. Such unique characteristics are present due to manufacturing variations. Silicon PUFs are still an active area of research since one needs to find functions that provide a high dependability on manufacturing variations but that are not sensitive to changing environmental conditions, e.g. temperature changes [REC04].

Silicon PUFs can be implemented using much less gates (in the order of some hundredth, see [REC04]) than cryptographic primitives like hash functions. In addition, silicon PUFs can have the property that it is even for the manufacturer difficult to produce two integrated circuits with identical PUFs. This might have advantages if the manufacturer is not regarded trusted and the material for key-value pairs shall already be applied at the time of manufacturing to save costs.

Using a silicon PUF instead of a list of key-value pairs in the tag is straightforward. In the manufacturing process or when the tag is attached to the product, the response of the PUF to a number of inputs is evaluated. Inputs and outputs form a list of key-value pairs. These pairs are given to the entity in charge of the tag, i.e. the tag owner here, and are stored there.

The use of PUFs in the context of RFID technology has been proposed by Ranasinghe et al. in [REC04] and by Tuyls and Batina in [TB06]. Whereas the latter focuses on off-line authentication, the former uses a PUF in combination with a challenge-response protocol similar to the one presented here. The difference between the proposal of Ranasinghe et al. and the approach presented here is the policy-based control of the use of key-value pairs. Only with such release policies, an attacker initiated depletion of key-value pairs can be prevented.

Summary

The presented *policy restricted key-value pair authentication* approach enables to reliably check the genuineness of products along the supply chain using inexpensive RFID tags. The tags do not require the implementation of a cryptographic primitive. In contrast to the track & trace approach, not just a plausibility check is provided. In addition, neither a fine-grained network of readers nor the keeping of a product history are required for the approach to work properly. In sum, the presented approach is inexpensive, feasible, and secure.

The approach is based on keeping a shared list of key-value pairs in the tag and at the respective tag owner. Usage of key-value pairs for secure authentication is restricted by application-area specific policies. These policies reserve sets of key-value pairs for parties acting in certain roles and control the release of pairs. These policies are the innovative part of the approach.

Release of key-value pairs is based on the role of the requestor, the extent of trust that the tag owner provides to the requestor, and the plausibility of the request. For instance, a reading party that identifies itself is provided more trust than an anonymous requestor since the former can be hold liable for its activities.

Constraining the release of key-value pairs counteracts depletion of these pairs so that attackers do not obtain the necessary information to become able to clone tags and therewith do not become able to create product counterfeits. The number of pairs that need to be kept on the tags can be kept rather low. This saves memory and therewith keeps tags inexpensive.

Instead of keeping a list of key-value pairs in the tags, a physical uncloneable function (PUF) can be implemented in the tags. This is a variant of the approach. However, the policies are still an indispensable part of the solution since the list of key-value pairs stored by the tag owner is still limited.

7.2 ID-Zone Architecture

In this section, a new RFID system framework called *ID-Zone Architecture* is introduced. Its goal is to address the issues regarding practicability that have been identified in the assembled system shown in the previous chapter, see section 6.5. The limit to use hash functions as only cryptographic primitive in tags persists.

In the privacy considerations performed in chapter 2, it has been shown that the user perception of privacy is the relevant aspect, not the privacy level that a system can actually provide. Based on that, the main idea is to obtain better system characteristics by relaxing the privacy requirements to some extent in areas where the privacy level is higher than practically needed.

This follows the general idea stated by Juels in [Jue04] to design RFID systems based on likely attack scenarios. But in the architecture presented here, the solution does not impair security and stays generic. This way, it is prevented that a false feeling of security is provided to the users and that applicability of the solution becomes limited.

Where appropriate, the architecture is organized in such a way that misbehavior of an entity is detectable and that the detected misbehavior can be sanctioned. This avoids the need of creating technical limitations that are difficult to implement and that worsen the system characteristics.

This section consists of several subsections. First, considerations regarding the actual requirements are taken. Based on these, the concept of "location zones" is introduced. After reasoning about device identifiers and certificates, basic considerations regarding tag identifiers are performed. With this background, an overview of the architecture is given and the procedure of tag identifier alterations in the architecture is explained. The subsequent main part of this section elaborates on the ID-Zone Architecture: The realization of the concept is presented along with proposals for the required protocols. Finally, an evaluation of the new architecture is performed.

7.2.1 Consideration of Requirements

In chapter 4, the concept of changing the identifier of an RFID tag was introduced to prevent unwanted recognition and tracking. If the identifier changes regularly in an unpredictable and forward secure manner, an attacker is no longer able to correlate the different identifiers so that he cannot decide whether obtained identifiers belong to the same tag (at a different time) or to different tags. Privacy is preserved this way, but a regular identifier change also introduces problems: The system does not scale as well as without an identifier change because each identifier change requires a transaction with the entity in charge of the tag. Additionally, chapter 5 showed, that the creation of changing pseudonyms for the entity in charge of the tag is a non-trivial and resource-consuming task.

In the considerations of the people's perception of privacy that took place in chapter 2, we learned that "total privacy" is not required. But the level of privacy needs to be adequate to the user's expectations. The question is whether we can lower the level of privacy in cases a high level is not required and therefore gain a more efficient overall RFID system. If one considers real-world scenarios in which people that carry objects to which RFID tags are affixed walk around, one can identify the situations in which the identifier of the tags should change and in which a change is not required.

If a person leaves a location like a supermarket and returns at a later time, the supermarket shall not be able to correlate the separate visits by default. Thus, the identifiers of RFID tags that are carried by the person need to be changed between the two visits.

If a person moves from one location to another, e.g. from one shop to another shop, it does not want that both shops are able to correlate the data they have independently collected. Thus, the identifiers of RFID tags need to be changed when the tags move from one location to another. The change needs to take place between the two locations in question so that none of the locations is able to observe the change, i.e. becomes able to get the old and the new identifier and thus be able to correlate the associated data. The physical space between the two locations can be regarded as a logical mix zone. This term "mix zone" has been introduced in [BS03] in the style of the term "mix network" that has been introduced by Chaum (see chapter 5 and [Cha81]).

While objects remain at a certain location, e.g. a supermarket or an office building, the identifiers of the RFID tags need not change. Usually, a local identification and tracking functionality is wanted, e.g. for goods in a storehouse. Tracking people by the items they carry in the scope of a certain location does not violate privacy much and can be regulated by privacy policies. There are fears by some privacy advocates that profiles of people's movements within a supermarket can be evaluated for optimizing product placement. But for most people, this is not a problem because there are also other techniques available to do so, e.g. CCTV cameras. In sum, the level of privacy can still be considered adequate if the identifiers of RFID tags do not change while remaining at a certain location. This cuts the number of required identifier changes significantly because most items do not move around.

Nevertheless, a change of identifiers makes sense for certain locations. For instance, in a supermarket scenario, nobody shall be able to scan all items that are available in the shelves, do the same the following day, and correlate the data. Such data could give a competitor useful information about stock turnover which can be regarded as corporate espionage. In other scenarios, e.g. for goods within a storeroom where nobody except the staff can enter, an identifier change needs not take place that often or is even not required at all. Thus, an identifier change at a certain location should be available optionally but not be mandatory. As the change shall be performed to obfuscate to extern entities and not to the owner of the location where items are present, the change is desired to take place in such a way that it can be performed by the owner of the location without requiring interaction with the tag owner. The changes can then take place more efficiently. Note that it should *not* be required to trust the owner of the location to behave correctly.

7.2.2 The Concept of Location Zones

In consequence of the considerations in the previous subsection, it makes sense to distinguish different locations. As there can be different logical locations (e.g. different offices on different floors of a building) at one geographical location, the term "location zone" is introduced: A location zone is defined as a physical space that belongs to the same entity, i.e. the person or organization that is in charge of the physical space. This entity will be called "location entity" in the following. Location zones do not overlap.

A zone identifier is assigned to each location zone. This identifier is unique for each location zone. The structure of the zone identifier is arbitrary in principle. For instance, identifiers based on "Global Location Numbers" that are assigned by GS1[1] can be used.

The author proposes to use zone identifiers that have a part for the country and a part that is directly linked to the geographical coordinates of the respective location zone. The reason is that the structure should be in such a way that directories of zone identifiers can be built easily. A field for the country is thus regarded useful so that each country becomes able to administer its own directory easily. Another requirement for zone identifiers is that their domain is large enough to ensure that attackers cannot perform brute force attacks in which all possible zone identifiers are created. One can append a random number to fulfill this requirement independently of the rest of the structure of the zone identifiers.

Zone identifiers need to have a prefix that denotes the kind of location zone to which a respective identifier belongs to. There are fixed location zones like buildings. Such location zones get zone identifiers with a prefix that will be denoted as "FL:" (for "fixed location") in the following description. Besides the fixed location zones, there are mobile location zones, e.g. the cargo area of a truck. Zone identifiers belonging to such location zones will be denoted with an "ML:" (for "mobile location") as prefix.

The concept of dividing space into different areas has already been introduced in [BS03] and inspired the approach shown here. In the cited work, a framework is presented that protects privacy by assigning pseudonyms to users. The pseudonyms are changed in so-called "mix zones", i.e. areas were many people pass through, so that devices or services in locations that the users visit are not able to identify or recognize the users. Whereas the focus of the article was on explaining the practical experience of the authors with the concept, here in this book chapter, the zone concept is transferred to RFID systems and will be extended and embedded into a complete architecture that is organized in a distributed manner.

7.2.3 Device Identifiers and Certificates

In the previous subsection, the physical space has been divided into different location zones. As a result, stationary RFID readers can be assigned to the location zone in which they are located. Each of such stationary RFID readers can be provided with a reader device identifier that starts with a prefix indicating that the reader is stationary. This prefix is denoted by "SD:" (for "stationary device") in the following. A reader device identifier should further contain the zone identifier of the location zone in which the reader is located. It can also contain the exact geographical location of the reader. To make the domain of the reader device identifiers large enough, the identifiers can be padded with a random number.

Besides stationary devices that have a fixed location within a certain location zone, there are also mobile RFID readers. The fact that they can be moved around and thus do not have a fixed location is expressed by reader device identifiers that start

[1] see http://www.gs1.org/glnrules/

with a prefix indicating a mobile device. The prefix is denoted by "MD:". Mobile devices are not assigned to location zones as they can be easily moved out of a certain zone. Thus, the reader device identifier of mobile devices does not link to a location zone. The identifier also does not contain geographical coordinates. Thus, the identifier can be made up using vendor information and a serial number and be padded with random values. This way, global uniqueness can be ensured.

Each location entity acts as a certification authority (CA) for the respective location zone and assigns certificates to the stationary readers within the respective location zone. The certificate acknowledges that the stationary reader with a certain reader device identifier belongs to the location zone and that it has correct geographical coordinates set.

The certificates of location entities themselves are signed by a higher level CA. That CA confirms that a location zone with a certain zone identifier belongs to the location entity and that information like the geographical coordinates is correct.

Using the certificates, each location entity and each stationary reader can prove its identity. The correctness of the information can easily be checked by independent parties so that one can rely on the information. The certification hierarchy is the basis of a secure and dependable RFID system.

RFID tags have identifiers, too. Within the architecture presented here, these identifiers consist of a zone identifier and a part that identifies the respective tag within the specified zone. This second part of an RFID tag identifier should be a random value so that valid identifiers cannot be guessed by an attacker. Like the other identifiers in the system, the domain of RFID tag identifiers and their parts should be that large that there are too many possible values for performing a brute force attack.

7.2.4 Basic Considerations Regarding Tag Identifiers

A main problem with the first approach of an overall RFID architecture that has been created within the chapters 4, 5, and 6 is that the tag identifiers do not have a structure any more but appear as random values. Linking these identifiers to the entity in charge of the tag requires pseudonymization infrastructures and is resource consuming. A solution that uses structured RFID tag identifiers would solve many problems of the presented architecture. On the other hand, structured identifiers enable recognition and tracking by constellation and have thus been avoided up to now.

However, using structured identifiers seems to be the only way to overcome the various problems that remain with the previously presented architecture. Therefore, as already introduced in the previous subsection, structured identifiers are used in the ID-Zone Architecture: The first part consists of a zone identifier; the second part consists of a random number to identify the RFID tag uniquely within the specified location zone.

The next question that arises is to which location zone an RFID tag should be assigned. Using the zone of the owner or the bearer of the tag is not a good idea since the zone identifier would provide a direct link to the owner or the bearer, respectively.

This would affect privacy since anonymity would not be possible. Alternatives would be to assign the location zone in which the RFID tag resides or to use any other location zone within the system. After careful consideration, using an arbitrary location zone appears to be the better solution. If an arbitrary location entity that is independent of the current location of the tag is in charge of the tag, this location entity can also provide the link to the owner and the bearer (see the previous chapter for the concept). This makes additional mixes or the like obsolete for most applications. In case a tag is moved out of the current zone without being noticed, an independent zone identifier that has nothing to do with the previous location zone, in which the tag has been, has the advantage that the previous tag location is not revealed. In addition, there are some other advantages of using an arbitrary location zone within the tag identifier. These advantages are regarding shared trust and will become clear later on after more information about the overall architecture has been provided to the reader.

As stated in the requirements, the identifier of an RFID tag shall change if it leaves a location zone and enters another location zone. But one cannot expect that RFID readers are located between each zone that can perform that change. Even if readers were present, one would need to take into account that an error could occur preventing the identifier change from being performed. One thus needs to find a way to perform an identifier change without explicit interaction when moving an RFID tag from one zone to another.

A straightforward solution is to perform the identifier modification as fast as possible after an RFID tag has left one location zone and has entered another zone. This means that the first RFID reader in the new location zone tries to perform the identifier change. The disadvantage of this approach is that the old identifier remains in use until the new location zone performs the change successfully. It would be better if the identifier change took place automatically without a special interaction as soon as the RFID tag enters another zone. This can be done if the identifier change has already been prepared so that the new zone only needs to activate the change.

The process of the preparation of the identifier change should not reveal the new identifier that the tag will have when it enters another location zone. This way, if many tags (more than one) enter a new location zone, an observer cannot determine which previous tag identifier belongs to which new tag identifier.

7.2.5 Architectural Overview

Based on the considerations in the previous subsections, a logical 3-layer architecture results. It is shown in figure 7.5. In the figure, the three layers for a single RFID tag with the identifier "ZoneB:Tag17" are shown.

This single RFID tag has a physical location where it currently resides. The corresponding (fixed) location zone is denoted by "FL:ZoneA" in the figure. Within this physical location zone, several RFID readers, either stationary or mobile ones, and many RFID tags can be present. As an example, a stationary reader with the identifier "SD:ZoneA:Reader1" is shown. This identifier indicates clearly that the reader

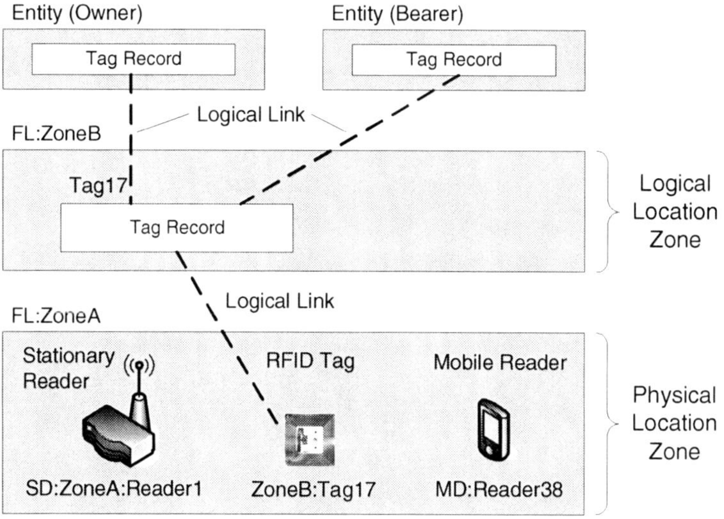

Fig. 7.5. Example of entities in the ID-Zone Architecture

belongs to the physical location zone. Besides this stationary reader, a mobile reader with the identifier "MD:Reader38" is shown in the figure. Mobile readers are independent of physical location zones since a mobile reader can easily be moved to another zone. This independency is expressed in the mobile reader identifier that is independent of a certain zone.

The example RFID tag shown in the figure has a tag identifier that has a first part which denotes another location zone. This location zone is the current logical location zone of that tag. That logical location zone is randomly chosen: Each physical location zone can act as logical location zone for tags that are physically somewhere else. In the example, the logical location zone of the tag is "FL:ZoneB".

The logical location zone manages the tag. It is responsible for changing the tag identifier and provides the link to the tag bearer and the tag owner which are the third layer of the architecture. Changing the tag bearer in case a tagged object is given away is also a task which is managed by the logical location zone of the tag.

Main tag states

Within the context of the presented layering, the identifier of each tag needs to change regularly to prevent illegitimate recognition and tracking. According to the considerations in the requirements subsection 7.2.1, tag identifiers shall change when an RFID tag moves from one location zone to another. In addition, it shall optionally be possible to change tag identifiers while the tags remain in the same physical location zone.

This is realized in the ID-Zone Architecture by introducing two main tag states, see figure 7.6. The tag state *LZ* is depicted on the left side. When the tag is in this

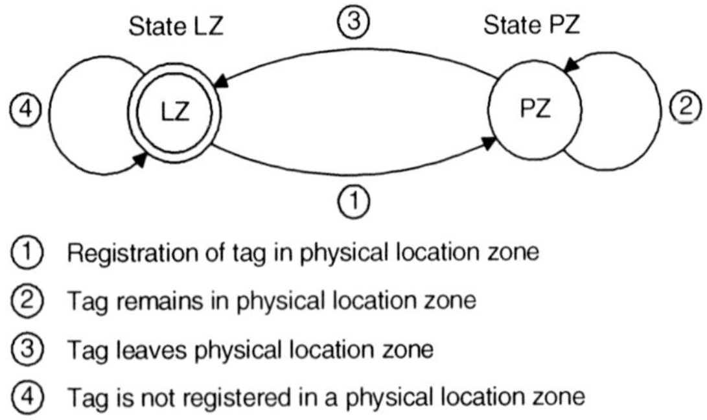

Fig. 7.6. Main tag states within the ID-Zone Architecture

state, it answers queries with a tag identifier that has as first part the identifier of the location zone that is currently responsible for the tag. The second part of the tag identifiers identifies the tag within that location zone.

When the RFID tag enters a physical location zone, it starts attempts to register to that zone. After successful registration, the tag enters state *PZ* via state transition 1. In this state, the tag answers queries with a tag identifier consisting of the identifier of the physical location zone in which it is registered and a second part for specifying the tag within that zone.

As all RFID tags that enter a physical location zone register there and thus get a tag identifier starting with the location zone identifier of that zone, the physical location zone identifier does not distinguish tags any more so that only the second part of the tag identifiers are relevant as distinguishing element.

If required for the prevention of corporate espionage or for other privacy reasons, the tag identifiers can optionally be changed while the tag remains in the physical location zone. The tag remains in the tag state *PZ*, see state transition 2 in the figure, the first part of the tag identifier is still the physical location zone identifier, and the second part of the tag identifier changes. This change is performed without requiring a complex, resource consuming message exchange.

When the tag leaves the physical location zone, the tag state reverts to *LZ*. This is state transition 3 in figure 7.6. This state transition does not require any involvement of the physical location zone like a deregistering process or the like. This ensures that nobody can hinder the state transition when the physical location zone in which the tag was registered is left.

Now back in the state *LZ*, the tag answers queries again with a tag identifier that has a logical location zone identifier as first part. But now the responsible logical location zone is another one than before the tag was registered to the physical location zone. Together with the previous registration in the physical zone, a handover from the old logical location zone to the one in use now has been prepared and becomes active now.

Thus, the physical location zone acted as a mix zone and the tag identifier has changed since the tag has been in the *LZ* state the last time. An observer who sees the new tag identifier has no clue which logical location zone the tag has belonged to earlier and what tag identifier the tag has had that time.

When the tag enters another physical location zone now and registers there, the described state transition 1 starts anew. As long as that does not happen, the tag remains in the *LZ* state, see state transition 4, and keeps the current tag identifier in tag queries.

7.2.6 Procedure of Tag Identifier Alterations

In the previous subsection, the two basic tag states a tag can enter within the ID-Zone Architecture have been described. In this subsection, the alterations of tag identifiers will be shown in greater detail to provide a better understanding of the occurring process.

Figure 7.7 depicts the principle of identifier alterations within the ID-Zone Architecture. As already explained, tag identifiers have two parts: a zone identifier and a random, zone-specific tag identifier. In the figure, only the zone identifiers are shown using the notation "zone identifier:". The second part is omitted there.

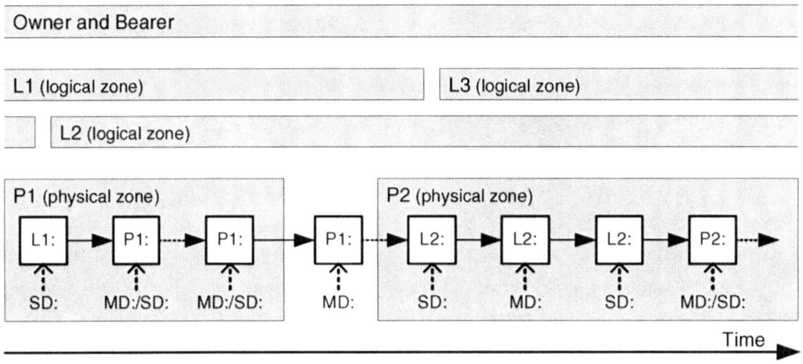

Fig. 7.7. Prefixes of tag identifiers over time

The figure again represents the three layers: The physical location zones in which the tags reside physically, the logical location zones in which a logical representation of the tags is kept, and the layer with owners and bearers. Again an example with a single RFID tag is shown to be able to explain what happens over time.

Imagine an RFID tag enters the physical location zone *P1*, see figure 7.7. When it is queried by a stationary reader (*SD:* depicting the prefix for a "stationary device") for the first time, it answers with a tag identifier which has the identifier of the current logical location zone of the tag as first part, *L1:* in the figure.

The physical location zone entity has no information about the RFID tag except its tag identifier. But with the first part *L1:*, the tag identifier gives a clue who to ask for more information: *L1:* denotes the logical location zone so that its entity can be contacted.

When that happens, the logical location zone entity *L1* performs several tasks. At first, it contacts another zone entity, in the figure the entity of the zone with the identifier *L2*. The location zone *L2* will take over the responsibilities of *L1* at a later time. The reason for this is that a regular change of the logical location zone is necessary to prevent the responsible location zone entity to create profiles of the tag movement.

Second, *L1* forwards requests of the location zone entity of the physical location zone *P1* regarding more information about the tag to bearer and owner. If the request is legitimate, *P1* will get the requested data via *L1*, otherwise not.

Third, *L1* sends some data to *P1*. This data is forwarded by *P1* to the RFID tag as soon as possible. The data contains information the tag can use to calculate the new tag identifier (i.e. zone identifier and zone-specific identifier) that it will have in *L2*. With the data, the tag can also check that the data is valid and comes from the logical zone entity that is currently in charge of the tag – invalid data is discarded.

Fourth, *L1* sends a key K_p to *P1*. This key can also be calculated by the tag so that *P1* and the tag then have a shared secret. The purpose of the key will be explained later.

Within the figure, it is assumed that the tag stays within the range of the reader for a long enough time so that all data destined for the tag can be sent to the tag within the same tag query process. The other scenario, which is also considered, will be explained later when the proceeding in *P2* is described.

If the tag received valid data, it uses it to calculate the new tag identifier that the tag will have in the logical location zone *L2*. This new identifier is stored in the tag but is not yet revealed to the outside. Using the data, the tag can also calculate K_p so that the current physical location zone and the tag have a shared secret in common now. The tag now enters another state that will come to play at the next tag query.

At the next tag query, the RFID tag does no longer answer with the tag identifier that is valid in its current logical location zone *L1*. Instead, it answers with a tag identifier that is only valid within the current physical location zone. Thus, the tag identifier now has *P1:* as first part, see the second square from the left in figure 7.7.

For this tag query and the subsequent ones in the same physical location zone, it is not relevant whether they are performed by a stationary or a mobile reading device: The tag answers in both cases with the same tag identifier.

A stationary reader within the current physical location zone *P1* can inform its location zone entity about the tag read. The location zone entity can recognize the tag using the information from the first tag query without being required to contact the logical location zone *L1* again. This significantly removes burden from the network infrastructure and enables a fast processing. A mobile reader device does not have any information about the tag except the tag identifier which has *P1:* as first part.

Thus, it has to ask the location zone entity *P1* for further information. Imagine a query for the price of an item in a supermarket: The mobile device of a customer can thus ask the shop directly without involving additional parts of the whole RFID system. Thus, such queries of mobile devices can be performed fast and efficiently. If the mobile reader asks for data regarding a tag that *P1* does not have or which *P1* is not willing to reveal, *P1* forwards the request regarding additional data to *L1* which itself forwards the request to tag bearer and tag owner who can ultimately take the decision to answer the request or to deny it.

Optionally, the second part of the tag identifier can be changed by the current physical location zone *P1* without involvement of the current responsible logical location zone *L1*. To do this, the established shared secret K_p is used in a manner that will be explained later. The possibility to change the tag identifier without involvement of the current logical location zone entity makes changes of the tag identifier an efficient task that does not put burden on the network infrastructure and on the current logical location zone entity. Note that the first part of the tag identifier, i.e. the current physical location zone identifier *P1:* remains the same and that only the second part is changed. But as all tags within the physical location zone get the same physical location zone identifier, tags can no longer be distinguished from each other using the first identifier part. Thus, the first identifier part cannot be abused for unwanted recognition and tracking by illegitimate parties.

After the RFID tag has left the physical location zone *P1*, the tag can be queried by mobile reading devices. The proceeding is the same as for queries within the zone: The reader obtains a tag identifier with *P1:* as first part and needs to ask the physical location zone entity for more information about the tag or the item to which the tag is affixed, respectively.

When the RFID tag is queried by a stationary reader of another physical location zone, the tag changes its state. It does not answer with a tag identifier starting with *P1:* any more so that, ideally, the new physical location zone entity does not learn where the tag has been before. Instead, the tag answers with a tag identifier with the new logical location zone *L2* as first part.

The physical location zone entity of *P2* now needs to contact the logical location zone *L2* for information regarding the tag. When the logical location zone entity of *L2* receives the request, the handover for the tag from *L1* to *L2* is finished and the entry for the tag in *L1* is no longer required. At that time, the handover to the future logical location zone *L3* of the tag is prepared.

The procedures within the physical location zone *P2* are in principle the same as for the physical location zone *P1*. But in figure 7.7 another potential scenario is depicted for *P2*: There it is assumed that the RFID tag does not stay in the read range of the first stationary reader long enough. Therewith, the tag cannot receive the data that has been created by the logical location zone entity *L2* and sent to *P2* which needs to forward it to the tag. The result is that in queries by mobile devices, the tag still answers with a tag identifier beginning with *L2:*. Such a first part of the tag identifier indicates that the tag is still new within the physical location zone. The mobile reading device needs to ask *L2* for information regarding the tag. Now the mobile device may forward the data from *L2* to the tag if it stays in range long enough. At latest, the next stationary reading device that queries the tag forwards the data that could not be transmitted to the tag in the first read attempt. Therewith, the preparation of the handover to *L3* is completed and the tag enters the state in which it answers with a tag identifier with the current physical location zone, here *P2:*, as first part.

Additional reads, either by mobile or by stationary reading devices, yield tag identifiers with *P2:* as first part. The stationary readers are part of the physical location zone so that the identifier can be directly linked to information regarding the tag. Mobile readers need to contact the physical location zone entity of *P2* for information regarding the tag. If that entity cannot answer the request, it can forward it to the current logical location zone entity of the tag, i.e. *L2*.

Description conclusion

The previous description presented an overview of what happens regarding tag identifiers within the ID-Zone Architecture and how the involved entities work together. The description shows that the separation of duties implements the concept of shared trust: Different entities have to work together to obtain a result. For example, a physical location zone entity does not communicate directly with tag bearer or tag owner, it even needs not know the identity of the tag bearer and the tag owner. Thus, the logical location zone entity works as mix. Changing the logical location zone entity for each tag regularly in a manner that cannot be anticipated gives protection against location zone entities that collect and forward data unauthorizedly.

The different involved entities, i.e. the ones involved directly as well as noninvolved observers, can check whether other entities work correctly. This way, misbehavior can be detected and therewith ultimately be sanctioned.

7.2.7 Elaboration of the ID-Zone Architecture

The previously presented overview of the ID-Zone Architecture describes an innovative concept. However, it surely reads as a wish list: Many parts appear to be desirable and make sense; but the presented description is just an overview which still misses ideas for implementation. This means that the protocols like the ones presented in chapter 4 with which the intended operation can be performed efficiently and securely are yet missing in the description. The following subsection will thus present concepts and protocols with which the described manner of operation within the RFID system can be implemented.

Preparation of tag identifier alteration during state change

Within the registration process of the tag in a physical location zone, the RFID tag shall be given the information into which logical location zone the handover shall take place when the tag leaves the physical location zone. An eavesdropping attacker shall not be able to learn which new logical location zone is in charge of the tag after the handover.

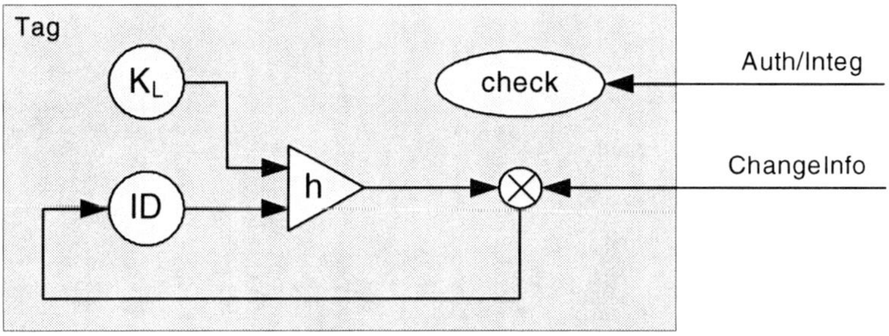

Fig. 7.8. Hidden change of tag identifier

The problem is thus to communicate information about the zone identifier of the new logical location zone without revealing it to observers. This would normally require an encrypted communication, but due to the goals of the architecture, a solution is required in which a hash function is sufficient.

A similar problem exists for the identifier update in the *Hash-based ID variation* approach, see section 4.8: The old tag identifier needs to be replaced by a new one without revealing the new one to an observer. The proposed solution is to submit a delta only. This has been implemented using a xor-operation: After submitting the delta, the new identifier is calculated using $NewIdentifier = OldIdentifier \oplus Delta$. As neither the old identifier nor the new identifier is revealed to the outside in the *Hash-based ID variation* approach, this solution is secure.

The situation here is somewhat different: The old identifier is possibly known to an observer. But the observer shall nevertheless not be able to calculate the new identifier using the old identifier and the data that is submitted to the tag.

A solution is possible using a shared secret of logical location zone entity and RFID tag by using hash functions niftily. The solution is shown in figure 7.8: Using the old identifier and the shared secret K_L, a hash value is calculated. This hash value is now used analogously to the old identifier in the *Hash-based ID variation* approach since it cannot be obtained by an observer. The tag can now obtain the new identifier by performing a xor-operation on the data received from the outside and the calculated hash value: $ID^* = ChangeInfo \oplus h(ID, K_L)$. The logical location zone entity has calculated that *ChangeInfo* before by calculating the same hash value and performing a xor-operation on the intended new identifier and that hash value: $ChangeInfo = ID^* \oplus h(ID, K_L)$. Note that the old identifier is only required as additional preimage of the hash operation if $h(K_L)$ shall be used for another purpose or multiple identifier changes shall be protected while K_L does not change.

It needs to be ensured that the *ChangeInfo* really originates from the logical location zone entity that is currently in charge of the tag and that an attacker is not able to alter that data undetectedly. Thus, an authentication of the sender needs to be performed as well as a check for data integrity. This can be implemented similarly as in the *Hash-based ID variation* approach by using a hash operation. Details on that will be given later.

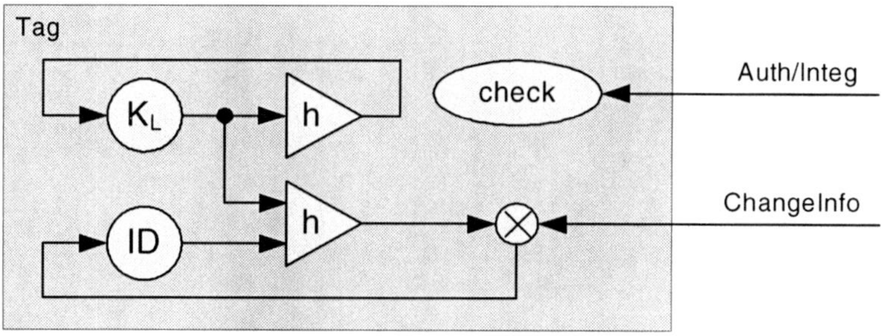

Fig. 7.9. Forward secrecy and hidden identifier change

When the location zone entity that is in charge of a tag changes, the secret K_L needs to be given to the new location zone entity so that it can prepare the next handover. The simplest solution would be to use a K_L remains constant.

But as long-term secrets should be avoided, a logical location zone entity can give away the hash value of its shared secret K_L instead. In this case, the RFID tag needs to perform the same operation as shown in figure 7.9. This way, forward secrecy can be ensured: If an attacker obtains a current shared secret K_L, he cannot conclude the former shared secrets $(K_L)_{-n}$.

Alteration of tag identifiers in physical location zones

The previous subsection described important aspects of the change of the tag identifier regarding location zones. In this subsection, the change of tag identifiers that occurs when a tag is registered to a physical location zone is described. In addition, the means by which tags detect in which physical location zone they reside is presented.

Reader broadcasts

As already mentioned, the RFID tags shall change their identifiers when they get into another location zone. Thus, the RFID tags need to be able to detect where, i.e. in which location zone, they are. To accomplish this, the stationary readers broadcast their zone identifier when they power-up tags. Using the zone identifier, the tags can detect whether they are in the same location zone as before or whether they have been moved into another location zone. Since the zone identifiers are fixed and do not change, one can easily detect whether stationary readers broadcast a correct zone identifier or something else. Misbehavior can thus be sanctioned.

As stated in subsection 7.2.1 about the requirements, even for RFID tags that remain within a certain location zone for a longer period of time, it makes sense to alter the tag identifiers from time to time. To accomplish this, the stationary readers also broadcast the current date and time. The tag uses the time information to trigger the modification of its identifier.

The location entity must obtain current time information via NTP [2] and distribute the time information to the stationary readers. The distribution can also be performed via NTP, but the location entity must be the only source for time information for stationary readers. This way, it is ensured that all stationary readers within a location zone have the same and exact time information.

The time information that is broadcasted by stationary RFID readers can of course be used by devices that do not belong to the RFID system, too. Devices that are not connected to a network and do not have a DCF77[3] antenna can obtain an accurate time this way, because stationary RFID readers can be expected to become ubiquitous in the future.

The stationary readers amend the broadcast with information denoting which part of the time information is used to trigger the alteration of RFID tag identifiers. For instance, a physical zone entity could define that the tag identifiers should change daily. This information together with the broadcasted time information defines an "epoch". Using the broadcasted epoch, the RFID tags are enabled to alter their tag identifiers independently from other extern action: If the epoch changes, the tag identifier is also changed without requiring a message exchange with the tag. This way, it can be ensured that alteration of tag identifiers really takes place.

Again, the correctness of the broadcasted information can be checked by any observer. This way, a location entity can be penalized if it broadcasts wrong information.

Alteration of tag identifiers based on reader broadcasts

Ideally, the tag identifiers are changed without a complex interaction between the RFID tags and the entities that are in charge of the tag. Only this way, the RFID system becomes well scalable and does not put too much burden on the infrastructure. Additionally, it needs to be assured that tag identifiers are really modified and that an identifier change is not hindered by a misbehaving entity or an attacker.

The stated requirements can be addressed by basing the alteration of tag identifiers on the reader broadcasts that have been introduced in the previous subsection. The principle is illustrated in figure 7.10.

Stationary readers broadcast the zone identifier of their zone and the current epoch to all tags around when powering tags. The tags can store such values in non-volatile memory (E and Z_P in the figure). As long as a tag is in the state LZ, in which

---

[2] Network Time Protocol, see RFC 1305

[3] DCF77 is a time signal transmitter located in Mainflingen, Germany, see http://www.ptb.de/de/org/4/44/pdf/dcf77.pdf

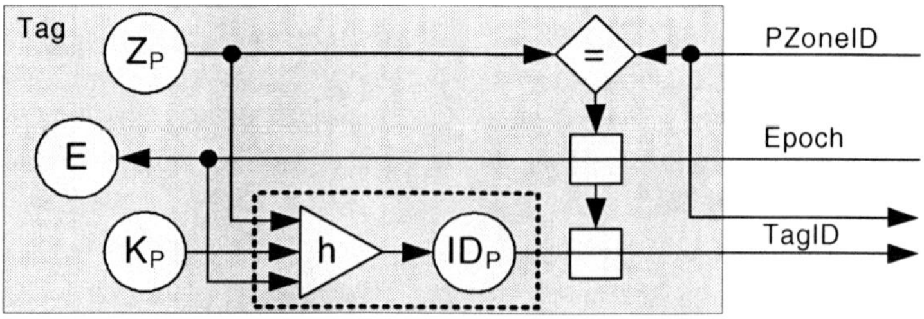

Fig. 7.10. Alteration of tag identifier based on reader broadcast

it answers with a tag identifier with a logical location zone identifier as first part, it saves the received physical location zone identifier and epoch values as Z_P and E in that non-volatile memory so that the last received values are always available to the tag.

In the preparation phase of the tag's state change, the tag obtains a secret K_P which is shared with the current physical location zone the tag resides in. K_P is stored in non-volatile memory, too. After the tag's state has changed, it answers with a tag identifier belonging to the current physical location zone it resides in, i.e. Z_P, as first part.

As long as the tag remains in that physical zone and thus keeps that state, the procedure upon a tag query is as follows: At a tag query of a stationary reader, the tag receives the physical zone identifier of that reader and gets the current epoch. As long as the received physical zone identifier equals the one stored in Z_P, the tag is still in the same physical location zone. Otherwise, the tag switches its state.

If the tag has not been moved out of the physical zone and the received physical zone identifier thus equals the one stored as Z_P, the received epoch is compared with E, and E is updated if the epoch has changed. The tag answers with a tag identifier with the physical zone identifier as first part and ID_P, i.e. a hash value based on Z_P, E, and the shared secret K_P, as second part.

As the hash-calculation takes some time which could delay the tag query, the tag can cache the calculated hash value in non-volatile memory. This is ID_P in the figure. Z_P and K_P remain static as long as the physical location zone does not change. Only the epoch E changes if a new epoch is received. As the epoch does not change very often, e.g. once per day, the second part of tag identifier can be obtained from the cache ID_P most of the time. Only when the epoch E changes, a new hash calculation, which updates the cache that contains the second part of the tag identifier ID_P, is required.

The physical location zone entity needs to perform the same kind of hash calculation for all the tags registered to the zone. Z_P is the location zone identifier of the physical zone and is thus a fixed value. For each registered tag, a shared secret K_P exists. The epoch E is based on current date and time and the update policy for tag identifiers within the physical location zone. The physical location zone entity has therewith all the information to calculate the current second part of the tag identifiers of the tags currently registered to the zone. The results of the calculation should be stored as an indexed field in the database table, which stores the currently registered tags within the physical location zone. If the epoch changes, the field needs to be updated by performing the hash calculation for all tags anew.

Using the indexed field in the database table of tags currently registered to the physical location zone, the physical location zone entity can quickly obtain the data associated to a tag after performing a tag query and obtaining the tag identifier: The tag identifier has the physical location zone Z_P as first part. Therewith, the physical location zone entity notices that the tag is currently registered within the zone. The second part of the tag identifier identifies the tag within the physical location zone. Data regarding the tag can be found in the database table containing registered tags by obtaining the row whose current tag identifier field equals the second part of the received tag identifier.

In contrast to stationary readers, mobile readers do not broadcast a zone identifier and an epoch value. The tag answers a query of a mobile reader with the zone identifier Z_P, the epoch E, and the cached hash value ID_P. Z_P and ID_P form the complete tag identifier, E denotes the epoch in which it is valid. Now the mobile reader has to contact the physical location zone entity denoted by Z_P to request more information about the tag. If the epoch E received from the tag matches the current epoch of the physical location zone, the physical location zone entity can recognize the tag using ID_P. If the epoch has changed, the physical location zone entity replies with the current epoch. The physical location zone entity does not calculate the tag identifiers of registered tags that would be valid in the provided epoch E since this is a resource consuming task that could be abused for a denial-of-service attack. The mobile reader now has to perform a second tag query in which it acts like a stationary reader: It sends Z_P and the current epoch to the tag. The tag then updates the stored epoch E with the current information, recalculates ID_P and provides the new value to the mobile reader. Now the mobile reader can query the physical location zone entity Z_P again with a valid second part ID_P of the tag identifier. This way, the mobile reader can obtain further information about the tag.

Altogether, the presented scheme is an efficient method for optionally changing tag identifiers while a tag remains within the same physical location zone. The current tag identifier is calculated based on the zone identifier of the physical location zone, the current broadcasted epoch, and the secret that the tag shares with the physical location zone entity. Thus, no explicit action is required for changing tag identifiers. Only in the case that the epoch changes, the tag needs to perform a hash calculation. Otherwise, queries can be answered without any delay out of the cache.

The scheme is secure since no harm can be done by querying the tag with faked values for the zone identifier and the epoch information. If the tag is queried with another zone identifier, the tag assumes that it has moved out of the physical zone and switches into the other tag state. This does not help an attacker. If the attacker only uses a faked epoch, the tag changes the stored epoch value and calculates a new hash to be used as second part of the tag identifier. A correct query with a valid epoch will reestablish correct information in the tag. Thus, an attacker cannot bring the tag into an invalid state using disallowed tag queries.

An interesting attack that can affect privacy would be to perform tag queries with future epochs. Therewith, an attacker could learn the identifiers that a tag will get in the future. But the identifiers are only valid as long as the tag does not leave the physical location zone. In other zones, the identifiers are completely different as they are based on the zone identifier. If the tag leaves the physical location zone and comes back at a later time, it will have another shared secret and therewith calculate other identifiers. There can thus only be a problem when the tag remains in the physical location zone. Performing the attack for many tags takes much effort for an attacker if he does not want his malign activities to be detected by observers. An attack performed carefully for few tags is surely possible. An attacker becomes therewith able to recognize these special tags in the special epochs for which the attack has been performed even when the tag identifiers have changed. But there is no means for an attacker to recognize the tags in arbitrary epochs. The attack has thus a very limited scope. The potential impact of such an attack is thus considered low so that the scheme still provides an appropriate level of privacy protection.

Overview of the complete protocol

After the introduction of some basic concepts of the ID-Zone Architecture in the previous subsections, a complete protocol overview shall be given in this subsection. The independent building blocks of the previous subsections are tied together and important missing elements for gaining a complete protocol are presented. The description will focus on the usual case to thus give an impression of the overall operation. Special cases like a changing epoch during a message exchange have been considered in the protocol design but are not presented in detail here.

The notation used is as follows: Capital letters denote variables; lower-case letters denote current values of variables. For being able to express the change of values over time, the values are indexed with a number that indicates a stage in time. Tag identifiers consisting of two parts will be denoted in the form Z/ID in which Z is the zone identifier as first part and ID the identifying number of the tag within the zone. For a better overview, values that have not changed and/or are currently not relevant are just denoted by "%" in the tables.

S	K_L	K_P	Z_L/ID_L	Z_P/ID_P	E
$L/\%$	k_{L1}	$\%$	z_{L1}/id_{L1}	$\%/\%$	$\%$

Table 7.5. Initial tag memory

Table 7.5 shows the tag memory when a logical location zone is directly in charge of the RFID tag. There is a variable S that stores the current tag state. It has a Boolean flag that can be either L or P which denotes whether the tag is currently registered to a physical location zone P or not. In the latter case that is currently given, the flag indicates with L that the tag is not registered to a physical location zone and that instead the current logical location zone of the tag is in charge of the tag directly.

The variable K_L contains the secret that the tag shares with the logical location zone. In the table, this secret currently has the value k_{LI}. The variable K_P contains the secret that the tag shares with the physical location zone in which the tag is registered to. As the tag is currently not registered to a physical location zone, the value of K_P is undefined.

The variable Z_L contains the zone identifier of the logical location zone currently in charge of the tag. The variable ID_L is the identifier of the tag within that zone. In the table, the two variables contain z_{LI}/id_{LI} so that the location zone in charge of the tag is z_{LI} and the identifier of the tag within that zone is id_{LI}.

When the tag is registered to a physical location zone, the variable Z_P contains the zone identifier of that physical location zone. The variable ID_P contains the tag identifier within that location zone. The variable E stores the last seen epoch of the physical location zone. The values of the three variables are currently undefined as the tag is not registered to a physical location zone.

In the current stage, the tag replies to the reader with the tag identifier z_{LI}/id_{LI} on tag queries. The interrogator thus needs to contact the logical location zone $L1$ with the zone identifier z_{LI} to get more information about the tag. The request is forwarded from the logical location zone to tag bearer and tag owner who return the requested data or deny the data divulgement. This answer is returned to the interrogator by the logical location zone entity. Thus, the logical location zone works as a mix so that the interrogator does not learn the identity of tag bearer and tag owner if the latter two do not reveal it.

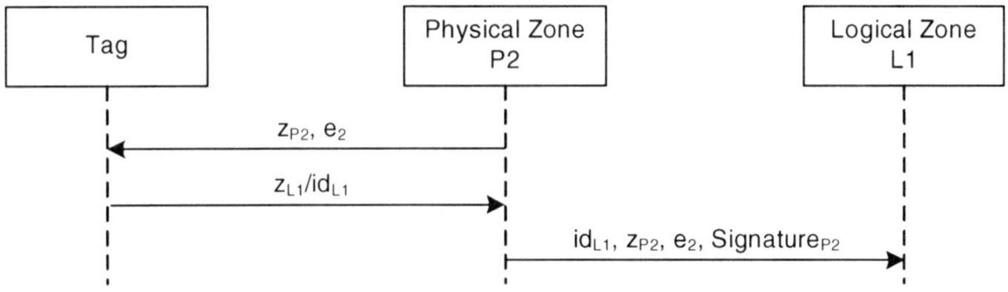

Fig. 7.11. Tag query by stationary reader with tag in L-state

The tag remains in this state and therewith the first stage of this description as long as it is only queried by mobile reading devices. But the situation changes when the tag is queried by a stationary reading device. This is the second stage in this description. The stationary reading device broadcasts the zone identifier of the physical location zone (it is denoted by z_{P2} in the example), the current date and time, and an epoch mask. This broadcast is the first message depicted in figure 7.11. The epoch mask is similar to the network mask in IP networks. Date and time and the epoch mask together provide the tag with the current epoch of the physical location zone. In the example, the current epoch value in this second stage is denoted by e_2.

The tag still answers with its tag identifier z_{LI}/id_{LI}, see the second message in figure 7.11. The physical location zone entity now contacts the logical location zone with the zone identifier z_{LI} to obtain additional information regarding the tag. The physical location zone entity sends the tag identifier id_{LI} so that the logical location zone entity is able to uniquely identify the tag within the zone. The physical location zone entity further sends its zone identifier z_{P2} and the current zone epoch e_2. The physical location zone entity signs the complete message with its private key so that integrity and authenticity of the message are assured.

The logical location zone entity can now check the signature and is then sure that the provided location zone identifier z_{P2} is valid. The zone identifier z_{P2} will later be used to link the reply message to the physical location zone. As the epoch e_2 is based on current date and time, the validity of that value can also be checked. If the physical location zone entity is sure that it has the correct date and time, it can sign the message including its own timestamp. If the physical location zone entity provided wrong information, the logical location zone entity would now act as a witness. If many logical location zones observe such a misbehavior, a legal entity supervising the complete infrastructure can sanction the physical location zone entity for providing wrong information.

The logical location zone entity can now contact tag bearer and tag owner to obtain additional information that has been requested by the physical location zone entity regarding the tag. In their decision whether information shall be released or not, tag bearer and tag owner should take into account that the original tag is perhaps not present in the physical location zone but that a mimicking device has provided the tag identifier z_{LI}/id_{LI} to the reader. One option here is to delay the release of additional information until the authenticity of the tag has been proven.

If not already done, the logical location zone can finish the handover from an old location zone to the current location zone. The presence of the valid tag identifier z_{LI}/id_{LI} "in the wild" proves that the original tag has revealed the identifier so that the previous logical location zone is no longer required for the tag.

The previous location zone should not free the allocated resources for the tag at once. If it receives an old tag identifier, i.e. a tag identifier of a tag whose handover is complete, then either a mimicking device is present in the system or the current logical location zone cheats by pretending that the handover is complete albeit it is not. If all location zones have a kind of reputation value that is managed by a legal entity supervising the complete RFID system, evidence of cheating that is witnessed by multiple other zones can decrease the reputation score and even lead to penalties. Accusation of cheating that is not acknowledged by other zones can in return decrease the reputation of the accuser. Note that there are no anonymous zones; only mobile reading devices can act anonymously in the system. As location zones prove their identity in the system using certificates, misbehavior can easily be attributed to the causer.

If not already done, the logical location zone entity now needs to prepare the handover of the tag responsibility from the logical location zone *L1* into another logical location zone *L3*. At first, the new logical location zone *L3* needs to be selected. This should be done amicably with tag bearer and tag owner to ensure that it is not possible that two malicious logical location zone entities take turns. In the new zone, the tag gets a new identifier id_{L3} that is unique within the location zone *L3*. The information regarding tag bearer and tag owner is copied to the new logical location zone. The new location zone further gets $k_{L3} = h(k_{L1})$ as shared secret that will be in common with the tag. The hash operation is performed to gain forward secrecy. Therewith, the new logical location zone is prepared.

The current logical location zone *L1* now obtains the *ChangeInfo* as already presented in one of the previous subsections. Therefore, $ChangeInfo = (z_{L3}, id_{L3}) \oplus h(z_{L1}, id_{L1}, k_{L1})$ is calculated. The first parenthesis denotes the bitwise concatenation of z_{L3} and id_{L3}. The hash value should have the same number of bits. This can be ensured by trimming some bits of the hash value if necessary.

The logical location zone also calculates $k_{P2} = h(k_{L1}, z_{P2}, e_2, ChangeInfo)$ and stores that value locally. Optionally, z_{L1} and id_{L1} can be used as additional preimages of the hash calculation. The value k_{P2} will be used as shared secret between physical location zone and tag and will also take part in ensuring the integrity of the reply message and for proving the validity of the message source to the tag.

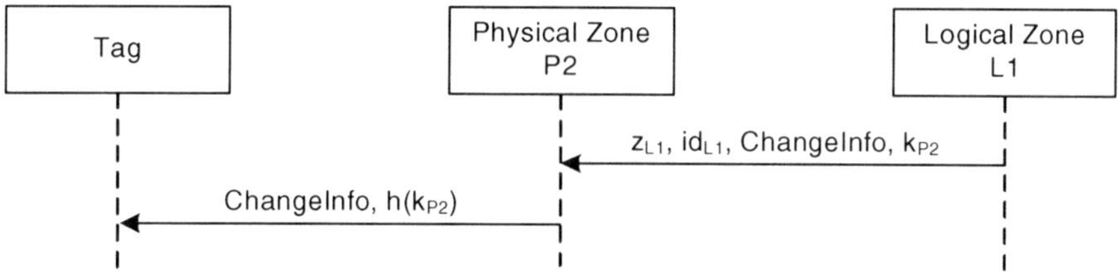

Fig. 7.12. Reply message of logical location zone entity

The calculated *ChangeInfo* and k_{P2} are then sent back to the physical location zone, see figure 7.12. In this message, the value id_{L1} together with the zone identifier z_{L1} of the message source identify the tag which the message belongs to. The physical location zone entity stores the received k_{P2} as shared secret that it will have in common with the tag after successful registration of the tag in the physical location zone.

The physical location zone entity calculates $id_{P2} = h(z_{P2}, e_2, k_{P2})$ and stores that value. The identifier id_{P2} is the second part of the tag identifier that the tag will have after successful registration in the physical location zone *P2* as long as the zone epoch does not change.

As depicted in table 7.6, the physical location zone has a zone identifier Z_P, here z_{P2}, and a current epoch E, here e_2, that are both valid for the complete location zone, i.e. for all devices and tags within that zone. For each tag, the logical location zone

Z_P *(per zone)*	E *(per zone)*	ID_P	K_P	Z_L/ID_L
z_{P2}	e_2	id_{P2}	k_{P2}	z_{LI}/id_{LI}

Table 7.6. Tag data in physical location zone

entity stores the identifier of the current logical location zone of the tag Z_L, here z_{LI}, the identifier ID_L of the tag within that logical location zone, here id_{LI}, the identifier ID_P of the tag within the physical location zone, here id_{P2}, and the secret K_P that is shared with the tag, here k_{P2}. Besides these values, the physical location zone entity can link arbitrary additional data to the tag.

The physical location zone entity calculates $h(k_{P2})$ and takes the next possible opportunity to send the received *ChangeInfo* and that hash value to the tag. This is the second message shown in figure 7.12. The shared secret k_{P2} is not sent in clear but in a hashed form so that the secret is not revealed to an eavesdropper.

The k_{P2} is on the one hand used as shared secret and on the other hand for authentication purposes as $h(k_{P2})$. An alternative would have been to use two different values that are sent from the logical location zone entity to the physical location zone entity. But the selected approach has the advantage that the values are bound to each other: This ensures that the physical location zone entity has received the correct shared secret if the tag receives the correct update information. This way it is ensured that the physical location zone is able to identify the tag by itself in the future as long as it remains in the physical location zone.

The tag evaluates the received *ChangeInfo* and $h(k_{P2})$: At first, the tag calculates $k_{P2tag} = h(k_{LI}, z_{P2}, e_2, ChangeInfo)$ by itself. The *ChangeInfo* is the value received; all other preimages are already present in the tag memory. The tag now compares the hash value $h(k_{P2tag})$ with the received hash value $h(k_{P2})$. If the two do not match, the received message is discarded and no further action besides sending an error notification to the physical location zone entity is taken. If the two values do match, the tag can be sure of several things: (a) The received *ChangeInfo* is correct and has not been altered due to a communication error or the action of an attacker. (b) The message originated from the logical location zone in charge of the tag because only the logical location zone entity is in possession of the shared secret k_{LI}. (c) The tag can be sure that the values z_{P2} and e_2 that it has received from the physical location zone entity are valid because they have been validated by the logical location zone entity. As the epoch e_2 contains date/time information, the tag is now in possession of a reliable and trusted current time. This might prove useful in a variety of application scenarios, e.g. in RFID tags combined with sensors to assign the correct time information to acquired sensor data.

If everything is fine so far, the tag stores the calculated k_{P2tag} in the tag memory as shared secret that it has in common with the physical location zone. Now the tag calculates $k_{L3} = h(k_{LI})$ and $(z_{L3}, id_{L3}) = ChangeInfo \oplus h(z_{LI}, id_{LI}, k_{LI})$ and changes the tag state into P. It is important that the two results of the calculations and the new tag state are stored in tag memory in a transactional manner so that either both results and the new state are stored correctly or the old values remain unaltered. In addition to the other operations, an authentication flag A is set in the state variable in the tag. Now the tag is registered to the physical location zone and the handover into the new logical location zone is prepared.

The tag can now calculate the second part of the tag identifier that the tag currently has in the physical location zone and cache that value: $id_{P2} = h(z_{P2}, e_2, k_{P2})$. As the tag is now registered within the physical location zone, it will no longer answer queries with the tag identifier z_{L1}/id_{L1} but instead with the tag identifier z_{P2}/id_{P2}.

S	K_L	K_P	Z_L/ID_L	Z_P/ID_P	E
P/A	k_{L3}	k_{P2}	z_{L3}/id_{L3}	z_{P2}/id_{P2}	e_2

Table 7.7. Tag memory after registration in physical location zone

After successful execution of the described operations, the tag memory has encountered a lot of changes. The resulting memory content is shown in table 7.7. The state variable S has the main state flag set to P which means that the tag is registered to a physical location zone. Another Boolean flag is set to A which means that the tag is in authentication mode. The variables Z_P/ID_P contain the current tag identifier of the tag. ID_P is only a cached value, it could also be calculated on the fly. The variable E contains the last seen epoch of the current physical location zone. The variables Z_L/ID_L contain the tag identifier that the tag will use after leaving the current physical location zone. Finally, the variable K_L contains the shared secret that the tag has in common with the logical location zone with the zone identifier Z_L.

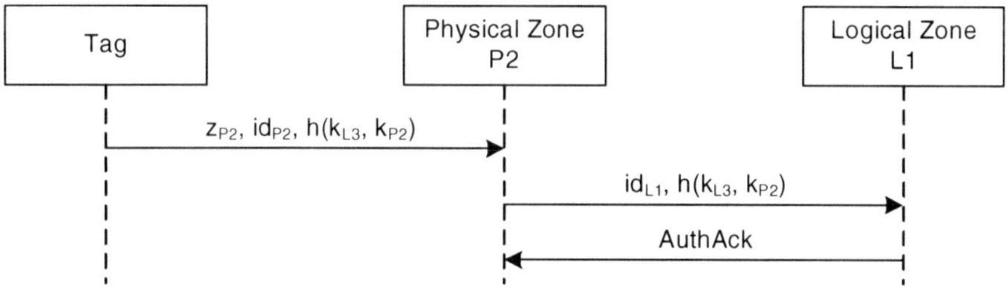

Fig. 7.13. Tag authentication

As long as the authentication mode flag is set, the tag does not only return the tag identifier z_{P2}/id_{P2} on queries but also the hash value $h(k_{L3}, k_{P2})$, see figure 7.13. Instead of the preimage k_{L3}, also k_{L1} could have been used if that value had still been available in the tag memory.

The hash value $h(k_{L3}, k_{P2})$ is forwarded to the logical location zone entity in charge of the tag. The logical location zone entity now calculates the same hash value $h(k_{L3}, k_{P2}) = h(h(k_{L1}), k_{P2})$. If the values match, the logical location zone can be sure that the original tag is the origin of that value and not a mimicking device.

The logical location zone entity learns further that the registration of the tag in the physical location zone has been successful. The logical location zone entity therewith also knows that the tag has obtained the epoch that has been validated by the logical location zone entity. This might be of interest for some extended application scenarios.

Finally, the logical location zone entity acknowledges the genuineness of the tag to the physical location zone using an *AuthAck* message, see figure 7.13. Therewith, both location zones can be sure that they did not correspond with a mimicking device.

Note that the authentication steps of the tag that are shown in figure 7.13 are not required for the operation of the protocol. When the tag publishes the new tag identifier z_{L3}/id_{L3} after leaving the current physical location zone, this is a sufficient proof that the original tag has successfully performed the handover into the responsibility of the new logical location zone *L3*. Prerequisite is of course that attackers cannot guess identifiers or have a noteworthy chance to try different identifiers by brute force successfully. For ensuring that this is not possible, the requirement has been made in subsection 7.2.2 that the domain of the identifiers is large enough.

S	K_L	K_P	Z_L/ID_L	Z_P/ID_P	E
$P/-$	k_{L3}	k_{P2}	z_{L3}/id_{L3}	z_{P2}/id_{P2n}	e_{2n}

Table 7.8. Tag memory after epoch change in physical location zone

In regular intervals, a physical location zone may change its epoch. In a new epoch, all tag identifiers of tags registered to the physical location zone shall change. To declare a new epoch, the physical location zone uses a new epoch mask location wide. As tags reply with new tag identifiers that are dependent on the new epoch, the physical location zone entity needs to change the tag identifiers in its tag database according to the new epoch. This means that the logical location zone entity calculates $id_{P2ni} = h(z_{P2}, e_{2n}, k_{P2i})$ for each tag i. In this formula, z_{P2} is the zone identifier of the physical location zone and e_{2n} is the current epoch. Both are the same for the whole location zone, i.e. for all stationary devices and tags in it. The variable k_{P2i} is the key that the entity shares with the tag. Of course, this key differs from tag to tag.

On tag queries, the zone identifier z_{P2} and the new epoch e_{2n} is broadcasted. If a tag receives an epoch value that differs from the previously seen one, it updates the second part of its physical zone tag identifier: $id_{P2n} = h(z_{P2}, e_{2n}, k_{P2})$. Thus, the tag answers queries with z_{P2}/id_{P2n} as tag identifier, see table 7.8.

If the authentication mode flag is still set, it is now cleared. This means that the tag only responds with the tag identifier z_{P2}/id_{P2n} in the future and not with the additional authentication hash value $h(k_{L3}, k_{P2})$. The reason for not sending the authentication hash value any more is that it does not change with a changing epoch so that it could be abused for unwanted recognition and tracking while the tag remains in the current physical location zone. It would not have been a good idea to also include the current epoch as additional preimage for the hash calculation because a faked epoch can easily be broadcasted to the tag by a mimicking device so that an attack with a partly chosen plaintext could occur.

The tag cannot distinguish whether it is queried by the physical location zone or by a mimicking device that broadcasts the zone identifier of the physical location zone and an arbitrarily faked epoch. The reason is that the physical location zone does not prove its identity to the tag. Technically, such a proof could have been implemented based on the shared secret K_P that tag and physical location zone have in common. But this would have caused a requirement for hash operations in each tag query. In contrast, as long as the tag remains registered to the physical location zone, the current solution does only require a single hash operation for the identifier update when the epoch changes and no hash operation at all in usual tag queries.

Although the broadcasted epoch can generally not be trusted, there is a single exception to that: As already explained, the epoch is sent to the logical location zone entity which checks the validity. If everything is fine, the epoch becomes part of the hash value used for authentication purposes. Therewith, the tag can be sure that the epoch value checked by the logical location zone is fine. Thus optionally, after registration in the physical location zone the tag may discard all queries with epochs that lie before the date/time that is given in that single trusted epoch value, i.e. e_2 in the previous description.

The tag stays registered to the physical location zone as long as it is only queried by stationary reading devices of that physical location zone or arbitrary mobile reading devices. If the tag is queried by a stationary reader of another physical location zone *P4* which can be detected by receiving a broadcast with the zone identifier z_{P4}, the tag answers with the tag identifier z_{L3}/id_{L3} that belongs to the new logical location zone. Thus, the new physical location zone entity does not learn in which physical location zone the tag has been before and which logical location zone entity has been responsible for the tag previously. This gives a good protection against unwanted tracking of movements. Now the tag is in the same state as in the beginning of this description and the protocol cycle starts anew.

Additional considerations

In this subsection, a few additional considerations of some relevant topics are presented. These are topics that did not fit into the previous description but that are regarded to be too important to omit.

Migration to new cryptographic primitives

A problem that has been highlighted in this book is the sustainability of approaches regarding security and privacy of RFID systems. Providing the possibility to use different cryptographic primitives usually is add odds with scalability or location privacy.

The ID-Zone Architecture has superior characteristics. In the identification phase, when the tag reveals a tag identifier, a responsible entity can be contacted without any cryptographic primitive involved and the tag can be identified uniquely by that responsible entity. The responsible entity can now have an entry in its database record about the tag that provides information which cryptographic primitive the tag uses. Therewith, different tags can use different cryptographic primitives without affecting scalability negatively by any means.

Concretely, the responsible logical location zone entity has information which hash function the tag uses. This entity can thus use the correct hash function and can also tell the currently responsible physical location zone which hash function to use. An attacker who overhears the communication between tag and reading device does not learn which hash function is used since hash output appears as output of a random number generator and no explicit information regarding the hash function is transmitted.

But note that the output of hash functions, i.e. the hash values, may have a different number of bits. For instance, there are hash functions that yield 128 bit, 160 bit, or 256 bit output. But as already stated in section 4.2, the size of the exchanged messages should always be the same for different tags so that is it not possible to perform tracking by constellation by performing traffic analysis. Therefore, if possible, hash functions that yield the same number of bits should be used. This means that in the migration process, not too many different output lengths should be allowed to make tracking by constellation practically infeasible. But this should not be a problem in practice.

In sum, as different hash functions can be used in parallel easily within the ID-Zone Architecture, a migration to new hash functions can be performed smoothly without problems. This makes the architecture sustainable.

Change of tag bearer

In section 6.3, the problem of determining the current tag bearer of a tag has been described. Comparing locations has been envisioned as solution: If tag and potential tag bearer are not near to each other for a reasonable span of time, one can conclude that he is not the bearer of the tag. In contrast, if tag and potential bearer are near to each other for a reasonable span of time and there has been no other person around, he is probably the tag bearer.

Based on these considerations, a solution for tag bearer determination can be implemented using the ID-Zone Architecture. The tag owner may not be the entity that administers who the current tag bearer is. First, the tag owner might have an interest in tracking the tag bearer without consent of the latter. Second, there are scenarios in which the identity of the tag bearer shall not become known to the tag owner, e.g. at price queries in a supermarket. But the current logical zone entity is a neutral third party and can thus administer the tag bearer status.

Each potential tag bearer needs a personal manager as presented in section 6.4. In that section, the use of a personal device as user interface to the personal manager has been proposed. Such a device is mandatory now and needs to be equipped with an RFID reader.

The envisioned procedure is then as follows: A user's personal device regularly tries to make the user become the tag bearer of all RFID tags in range of the user. The current location zone entity of each tag records such requests. Thus, the location zone entity will regularly get requests of all users near the respective tag. Based on this information, the entity decides whether the current tag bearer changes or not and performs the change if necessary.

Imagine for example a library book tagged with an RFID tag. As long as the book is in the library, the library is the tag owner. There is no tag bearer or the library is the tag bearer, respectively. When the library lends out the book to a person called e.g. Sam, Sam becomes tag bearer. This change can be triggered at checkout as the latter is performed in some special user action.

Sam can move around with the book. His privacy is protected since he is registered as tag bearer. This means that all tag queries need to be admitted by his personal manager before the tag can be identified. As all tag queries by any readers are passed to the personal manager, Sam can use the book with the tag in the same way as if the book was owned by him.

The personal device that Sam carries around informs the current location zone entity of the tag regularly that he is still near the tag. When Sam goes for a walk with his girlfriend Emily with the book near to both of them, her personal device sends requests for becoming tag bearer to the location zone entity, too. As long as the location zone entity receives information that both people are near the tag, the tag bearer does not change. But if Sam gives the book to Emily and goes somewhere else, Emily's personal device will be the only one to send requests to the location zone entity. The latter will thus perform a change of the tag bearer. After that, Sam is no longer the tag bearer but Emily is. As the tag is now under her control, she can protect her privacy effectively.

Note that the presented procedure of determination of the tag bearer is only an overview. In practice, additional data would be used within the process. For instance, the relationship between physical location and tag owner and potential tag bearer can be taken into account. The value of an item, its portability, and its likelihood of movement can also be considered for adjusting the automatism of tag bearer determination.

Policies should exist, how long information about the requests of nearby users is kept by location zone entities. Such information should be regarded as ephemeral data and therewith be deleted if no longer required, e.g. after a few minutes. Furthermore, policies should be created for cases in which nobody requests to be the tag bearer of a tag. The respective items could have been set down somewhere or could have been lost. In such a case, the location zone entity should inform the current tag bearer. If the latter's personal manager cannot prove that the current tag bearer is still in possession of the respective item (e.g. it is located in physical space owned by the tag bearer), the tag owner should be set to be the tag bearer.

7.2.8 Evaluation

The presented ID-Zone Architecture consists of its core concepts and the protocols that implement these concepts. The components are independent of each other. For instance, one could replace one protocol by another one. This would change the overall system properties. Instead of evaluating each component independently of the other ones, the whole system including all the presented components in the presented implementation is evaluated in the following. Note that not all the resulting characteristics derive from the design of the architecture but to a large extent from the chosen protocol implementations.

SECURITY: The ID-Zone Architecture has explicitly been designed to address the goals regarding security and privacy (see section 3.3) in an economic manner. Thus, the architecture meets the goals well.

Data security is maintained by keeping data associated to the tag within the back-end infrastructure, i.e. at tag owner and tag bearer. The current logical location zone entity temporarily stores some data required for system operation, e.g. the tag's current owner and bearer and data required for tag bearer determination. As the current location zone entity is freely chosen and does not get to know additional data, that entity can be regarded as neutral party. If it is ensured by using encrypted communication that the location zone entities cannot overhear application specific communication, one can assume that the incentive for performing malicious activities is low. Penalties set by law and mandating and controlling the use of software that has been certified by independent supervision authorities can corroborate this assumption. The current physical zone entity does not get any data that could be abused so that no data security issues occur.

The prevention of product counterfeiting by impeding tags to be cloned is addressed in the architecture, too: Tags keep an inner state that is never revealed to the outside world. By using protocols that prove the validity of the tag state to the tag owner, genuineness of a tag is assured. In the concrete implementation, an authentication mode of a tag has been provided. In this mode, a tag reveals a hash value based on the tag's inner state. This authenticates the tag.

By only evaluating data after the authenticity of tags has been proved, infiltrating false data by an attacker into the system can be prevented. Injecting malformed or invalid protocol messages by an attacker does not lead to harm due to protocol design. Access to application specific tag data can be controlled using arbitrary access control schemes. As these are implemented at the tag owners and the tag bearers where many resources are available, the access control schemes can be implemented fine-grained and flexibly. The stated measures thus prevent illegitimate access to the system effectively.

Unwanted recognition and tracking is prevented by modifying tag identifiers regularly. First, the tag identifier changes when a tag leaves the physical location zone which it is currently registered to. Second, the physical location zone can change the identifiers of tags that are currently registered to the respective zone regularly based on the announcement of the current time and the epoch mask. An attacker could try to inhibit registration in a physical location zone, e.g. by intercepting messages to the tag. This would prevent identifier modification and would allow recognition and tracking. But for such attacks, an attacker must be close to the tag. If an attacker has such possibility, he can perform simpler means of recognition and tracking. Furthermore, other readers querying the tag would notice that the tag does not become registered to the physical location zone which raises suspicion.

Attacks that lead to a denial-of-service are prevented by protocol design: The employed protocols are able to cope with attacks like message interception or replay attacks. Hence, tag and backend infrastructure cannot be brought out-of-sync which would disturb proper system operation.

Other advantageous additional properties are provided by the architecture, too. Forward secrecy in identifier modification across zones is provided by updating the internal tag identifier and the shared secret using hash functions. In contrast, identifier modifications performed by a physical location zone entity are not done forward-securely. But this in not required as only little information can be revealed by an attack. There is practically no incentive to perform such an attack. The ID-Zone Architecture with the currently used protocols does not share any secrets among several tags. Thus, the incentive for attackers to perform elaborate physical attacks is very low.

RESOURCES: Resource consumption is to a large extent dependent on the protocols used to realize the concepts of the ID-Zone Architecture. One needs to distinguish the resource consumption in different states and processes. For instance, the resource consumption for tag queries while the tag is registered to a physical location zone is different to the one for the registration process in such a zone.

As resources in the tags and communication between tags and readers are the most relevant ones, the following discussion deals with these. A tag needs memory for storing the current tag state, two tag identifiers consisting of a zone identifier and an identifier valid in that zone, the currently announced epoch, and two shared keys. This can be regarded as very efficient in proportion to the features the architecture provides.

Tag queries require a single message exchange: Worst case is a stationary reader that broadcasts its physical zone identifier and announces the current epoch in that zone. This only needs to be done once for querying all the tags in the reader's read range. Each tag answers with its current tag identifier if that identifier is not already known to the reader from processing the anticollision protocol. Additional messages are only required for special purposes. On the one hand, this is the acknowledgement of the registration of a tag in a physical location zone resulting in one message from reader to tag. It contains the *ChangeInfo* and a hash value for providing backend authentication and message integrity. On the other hand, there are response messages from the tag that do not only contain the tag identifier. These messages are sent from tag to reader as long as the tag is in authentication mode and authenticate the tag. A hash value is added for this purpose.

Several cases can be distinguished regarding computation. As long as the tag is in state *LZ*, see figure 7.6, no computation needs to be performed on a tag query. The tag simply reveals its stored tag identifier. If the tag is in state *PZ*, the tag needs to perform one hash calculation in case the epoch announced by the reader is new. But as long as the epoch does not change, no calculation at all is required. Fortunately, this is the most probable case since most items do not move around and the epoch does not change very often.

Just like additional messages, additional communication is required for special purposes. For performing tag registration in a physical location zone, the tag needs to perform four hash calculations after receiving the acknowledgement from the logical location zone entity via the reader. For the creation of an authentication message, the tag needs to perform one hash calculation.

In sum, the only really resource consuming task is performing the registration to the physical location zone. This registration also prepares the tag identifier modification taking place when the tag leaves the physical zone again. For queries of tags registered to a physical location zone, no calculation at all is required as long as the epoch does not change.

SCALABILITY: The presented ID-Zone Architecture is organized in a completely distributed manner: There are no central authorities or special purpose services. Thus, there are no bottlenecks. Tag owners and tag bearers administer their own tags. Physical location zone entities administer the tags within their physical space. The load of logical location zone entities is shared among all existing entities.

As the load is shared, the databases in the system do not get very large. The burden on the infrastructure is kept low by delegating the ability to recognize and track the tags that are registered to a physical location zone to the respective physical location zone entity. This matches well with the practical demand as one can expect that the local traffic, e.g. for inventory control, will be high. The architecture is also well scalable regarding computation: The backend entities only need to perform database queries and hash calculations. The required number of such calculations is low compared to other usual backend operations like encrypting communication.

SUSTAINABILITY: It has already been explained that the ID-Zone Architecture is able to use different hash functions simultaneously. This provides a migration path towards new cryptographic primitives if the security characteristics of the old ones do not fit the requirements any more.

The impact of a successful attack against a single tag is limited to that single tag since no secrets are shared among tags. This way, the incentive for such attacks is low. Thus, it does not pose a problem that a broken cryptographic primitive does not only affect privacy but provides an attacker control over the respective tag as well.

PERFORMANCE: The most important performance criterion is the speed of tag queries. The latter is dependent on the message exchanges and computations that need to be performed. As already discussed, tag queries require a single message exchange. Either no hash operation is required at all in the tags or one hash operation is needed if the announced epoch changes. Thus, the tag identifier can be instantly delivered without delay most of the times. After announcing a new epoch, all tags in read range can perform the required hash operations in parallel. This leads to a delay of a single hash operation. The tag identifier can be sent afterwards without the need for additional computation. As no more data than in a basic tag query is required in the messages, the tag queries can be performed with maximum efficiency.

As stated in the paragraph regarding resources, the computationally most expensive operation is performing the registration of the tag to a physical location zone entity and thereby preparing the identifier modification for leaving the zone. Only a single message from backend to tag is required, but the tag needs to perform four hash operations. This high computational effort is not a problem in practice because it does not result in delays: Until the calculations are finished, the tag can use its logical location zone identifier. The calculations can be performed in the background; the tag just needs to be powered by a reader. If the tag caches intermediate results in non-volatile memory, the operations can even be resumed after power loss without requiring starting over again.

HANDLING/PRACTICABILITY: Zone transfers and identifier modification takes place automatically without requiring user attention. The architecture only employs the wireless RFID communication and no alternative channels. This makes handling easy.

Calculations need not be performed in a single step: Interrupted processes can be resumed if the interim results have been cached in non-volatile memory. Further, the tags need not stay online while the responsible backend entity is queried. This is a huge advantage compared to many other protocols. The ability to perform precalculation and the independence of the messages in the employed protocols results in a form of resilience of the protocols regarding sudden power loss, e.g. caused by user movement.

All these are very good characteristics: As a result, a user does not need to consider the protocol operations; they just work in the background. As the protocol operation is hardly dependent on special circumstances, e.g. tags staying in the range of a reader until all message exchanges are completed, careless user action does not hinder protocol operation.

However, there is also a yet unresolved issue for practical application: The presented protocols require the involved entities to be available. There are no failover mechanisms implemented that handle failure of system components.

For instance, once the logical location zone, to which a future handover shall take place, is selected and the tag programmed, this selection cannot be changed any more. If the respective logical location zone entity is not available when the handover shall be performed, the current logical location zone must keep the responsibility for the tag and try again later. During that time, the tag cannot be identified since the reading entity cannot contact the logical location zone entity that the tag denotes in its tag identifier.

To resolve the availability issue, one could introduce redundancy. If a tag always had two tag identifiers, a reading entity could switch to the entity denoted in the second tag identifier if the primary entity was not available. The drawback of this approach is that the required resources double. On the other hand, not only the availability problem is solved, but the two zone entities currently in charge of the tag can supervise each other. This confines the ability for single entities to disturb proper system operation even more.

UNIVERSALITY/SCOPE: The ID-Zone Architecture is a generic framework. It is independent of certain applications. It provides an infrastructure for tag identification and authentication and enables to perform identifier modification to thus gain location privacy. Tag owner, tag bearer, and reading entity remain anonymous to each other. Any additional application-specific functionality can be built on top. For instance, arbitrary data can be assigned to tags.

The ID-Zone Architecture is well scalable so that it can be used on a global scale. The possibility to delegate the ability to identify a tag to the physical location zone in which the tag resides fits the practical requirements well and removes much burden from the infrastructure.

The architecture supports the extensions of classic RFID systems introduced in the previous chapter. The newly introduced entities enable inter-organizational systems while enabling a flexible data sharing. All these desirable properties are present in the ID-Zone Architecture. Thus, it provides the foundation for building truly open RFID systems.

7.2.9 ID-Zone Architecture Summary

In this section, the ID-Zone Architecture has been presented. Its goal is to provide security and privacy in large-scale RFID systems supporting inter-organizational data sharing which is necessary in practice. To achieve this, new concepts have been introduced and afterwards been implemented using appropriate protocols.

Based on the observation that identifier modification on each tag read is not efficient, alternative models have been considered. The discussion of privacy in chapter 2 showed that the perceived privacy level is the relevant one. This led to the partition of the physical space into so-called physical location zones. For obtaining location privacy, an identifier modification when passing from one zone to another is sufficient.

That new zone concept alone is not sufficient: First, it needs to be ensured that the identifier modification really takes place. Second, a reader in a new zone should not get to know the previous location of the tag, i.e. any information regarding the old zone. Third, zone entities and readers cannot be fully trusted. Thus, principles like shared trust are important. Fourth, the approach shall be flexible and generic. All these issues, among others, need to be addressed in a scalable manner without requiring many resources or affecting performance negatively.

The solution to these requirements has been found in the concepts of distinguishing physical and logical location zones as well as using different tag states: Tags register to physical location zones. Therewith, the ability to identify tags is delegated to the respective physical location zone. This fits the practical requirements well. Furthermore, the identifier modification that will take place when the tag leaves the zone is prepared. Proper operation of the scheme is ensured by mutual supervision of the involved entities and of extern entities. This saves resources and leads to high efficiency since otherwise more complex technical measures would have needed to be taken.

While a tag is registered to a physical location zone, the respective physical location zone entity can trigger an identifier modification in the tags. Instead of communicating with each tag separately, the modification is implemented by using the new concept of time/epoch announcements. After reception of a new epoch, tags can calculate their new tag identifiers without interaction of a reader. This makes the identifier changes efficient and reliable.

Protocols for the realization of the new concepts have also been introduced in this section. This has been done based on the experience gained in the implementation of other RFID protocols. The result is secure protocols that do not require much computation and are optimized for supporting advanced features like precalculation.

The ID-Zone Architecture with its implementation based on the proposed protocols forms a comprehensive and generic solution. The evaluation attested excellent characteristics to it.

7.3 Summary

In this chapter, advanced solutions for RFID systems have been introduced. The challenge has been to improve some of the solutions presented before. The previous chapters put the focus of the considerations on security and privacy mainly. In this chapter, the practicability of solutions has been added as a primary goal. In addition, the possibility for some minor trade-offs regarding privacy protection in order to gain much better economic efficiency has been exploited.

The considerations started with the *Triggered hash chain* approach as an alternative to the *Hash-based ID variation* scheme presented in chapter 4. Both protocols provide identification, authentication, and identifier modification. The approach presented in this chapter is more efficient and more elegant. Furthermore, it allows tags to be offline while the reading entity communicates with the responsible backend entity. This is a huge improvement regarding practicability.

The second presented approach is *policy restricted key-value pair authentication*. It addresses a major problem in supply chains: product counterfeiting. The core innovation is to prevent depletion of authentication data using policies tailored to the application area. In contrast to approaches based on track & trace, it provides strong authentication and does not need a fine-grained infrastructure of RFID readers. Such characteristics normally require the implementation of a cryptographic primitive in the tags, but the presented approach can cope without. This makes it comparatively inexpensive. As the approach is also feasible and secure, it is an economically interesting solution to the counterfeiting problem.

Finally, a new overall RFID system architecture called *ID-Zone Architecture* has been presented. The goal has been to find a solution for large-scale RFID systems that do not only provide security and protects privacy but also is technically feasible and has good economic characteristics. The functionality to enable inter-organizational data sharing has also been regarded as a required functionality.

The ID-Zone Architecture introduces a variety of new concepts with advantageous characteristics fitting the practical requirements well. The required system extensions that have been introduced in the previous chapter, e.g. taking the tag bearer into account, are respected, too. The concepts have not only been proposed but have also been implemented using appropriate protocols. Altogether, the ID-Zone Architecture enables building scalable and efficient RFID systems on a global, inter-organizational scope without neglecting security and privacy.

Of course, before the concepts and architectures presented in this book can be used in practice, still a lot of work needs to be done. The promising ID-Zone Architecture needs refinement. Considerations regarding reliability and an in-depth security analysis need to be performed. The architecture has to be applied to a variety of scenarios and the impact of such a system on society needs to be studied. A practical implementation is of course also required.

List of Figures

List of Tables

References

[AAC+03] Dakshi Agrawal, Bruce Archambeault, Suresh Chari, Pankaj Rohatgi, and Josyula R. Rao, *Advances in Side-Channel Cryptanalysis, Electromagnetic Analysis and Template Attacks*, Cryptobytes, RSA Laboratories vol. 6 (2003), no. 1, pp. 20–32.

[ABKL91] Martin Abadi, Michael Burrows, Charlie Kaufman, and Butler W. Lampson, *Authentication and Delegation with Smart-cards*, Theoretical Aspects of Computer Software, pp. 326–345, 1991.

[AC03] Anita L. Allen-Castellito, *Why privacy isn't everything: Feminist reflections on personal accountability*, Rowman and Littlefield Publishers, June 2003.

[Acq02] Alessandro Acquisti, *Protecting Privacy with Economics: Economic Incentives for Preventive Technologies in Ubiquitous Computing Environments*, Workshop on Socially-informed Design of Privacy-enhancing Solutions, Fourth International Conference on Ubiquitous Computing (UbiComp 2002), 2002.

[Aho01] Jari Ahola, *Ambient Intelligence*, ECRIM News (2001), no. 47.

[Aho02] Jari Ahola, *Ambient Intelligence: Plenty of Challenges by 2010*, EDBT '02: Proceedings of the 8th International Conference on Extending Database Technology, Springer, p. 14, 2002.

[AO05a] Gildas Avoine and Philippe Oechslin, *RFID Traceability: A Multilayer Problem*, Financial Cryptography – FC'05 (Andrew Patrick and Moti Yung, eds.), Lecture Notes in Computer Science (LNCS), vol. 3570, IFCA, Springer, pp. 125–140, February/March 2005.

[AO05b] Gildas Avoine and Philippe Oechslin, *A Scalable and Provably Secure Hash Based RFID Protocol*, International Workshop on Pervasive Computing and Communication Security – PerSec 2005, IEEE Computer Society Press, pp. 110–114, March 2005.

[APCC94] Australian Privacy Charter Council, *Australian Privacy Charter*, 1994.

[Avo05a] Gildas Avoine, *Cryptography in Radio Frequency Identification and Fair Exchange Protocols*, Ph.D. thesis, EPFL, Lausanne, Switzerland, December 2005.

[Avo05b] Gildas Avoine, *Radio Frequency Identification: Adversary Model and Attacks on Existing Protocols*, Technical report, Swiss Federal Institute of Technology, Lausanne, Switzerland, September 2005.

[Avo07] Gildas Avoine, *Bibliography on Security and Privacy in RFID Systems*, available online at http://lasecwww.epfl.ch/~gavoine/rfid/, 2007.

[BfD05] BfD, *Tätigkeitsbericht 2001 und 2002 des Bundesbeauftragten für Datenschutz*, August 2005.

[BFHF03] James Brusey, Christian Floerkemeier, Mark Harrison, and Martyn Fletcher, *Reasoning About Uncertainty in Location Identification with RFID*, Workshop on Reasoning with Uncertainty in Robotics at IJCAI, August 2003.

[BFK00] Oliver Berthold, Hannes Federrath, and Marit Köhntopp, *Project "Anonymity and Unobservability in the Internet"*, Workshop on Freedom and Privacy by Design / CFP 2000, pp. 57–65, April 2000.

[BM87] Jr. Barrington Moore, *Privacy: Studies in Social and Cultural History*, American Journal of Sociology vol. 92 (1987), no. 5, pp. 1232–1233.

[BMBF07] Bundesministerium für Bildung und Forschung [Federal Ministry for Education and Research, Germany] *RFID-Studie 2007 – Technologieintegrierte Datensicherheit bei RFID-Systemen [RFID Study 2007 – Technology-integrated data security in RFID systems]*, http://www.tzi.de/fileadmin/resources/publikationen/news/RFID-Studie_Final.pdf, June 2007.

[Bri99] David Brin, *The Transparent Society: Will Technology Force us to Choose Between Privacy and Freedom?*, Perseus Books Group, June 1999.

[BS03] Alastair R. Beresford and Frank Stajano, *Location Privacy in Pervasive Computing*, IEEE Pervasive Computing vol. 2 (2003), no. 1, pp. 46–55.

[BSI05] Bundesamt für Sicherheit in der Informationstechnik [Federal Office for Information Security], *Risiken und Chancen des Einsatzes von RFID-Systemen (RIKCHA) [i.e. Risks and prospects of the use of RFID systems] - Trends und Entwicklungen in Technologien, Anwendungen und Sicherheit*, 2005.

[BSI06] Bundesamt für Sicherheit in der Informationstechnik [Federal Office for Information Security], *Pervasive Computing, Trends and Impacts*, http://www.bsi.bund.de/literat/studien/percenta/Percenta_elay.pdf, 2006.

[BSNP95] Shahram Bakhtiari, Reihaneh Safavi-Naini, and Josef Pieprzyk, *Keyed Hash Functions*, Cryptography: Policy and Algorithms, pp. 201–214, 1995.

[Bur97] Herbert Burkert, *Privacy-enhancing technologies: typology, critique, vision*, Technology and Privacy: The New Landscape, pp. 125–142, MIT Press, Cambridge, MA, USA, 1997.

[Cha81] David L. Chaum, *Untraceable electronic mail, return addresses, and digital pseudonyms*, Communications of the ACM vol. 24 (1981), no. 2, pp. 84–90.

[Cla01] Roger Clarke, *Introduction to Information Security*, http://www.anu.edu.au/people/Roger.Clarke/EC/IntroSecy.html, February 2001.

[CLL05] Eun Young Choi, Su Mi Lee, and Dong Hoon Lee, *Efficient RFID Authentication protocol for Ubiquitous Computing Environment*, International Workshop on Security in Ubiquitous Computing Systems – secubiq 2005 (Tomoya Enokido, Lu Yan, Bin Xiao, Daeyoung Kim, Yuanshun Dai, and Laurence Yang, eds.), Lecture Notes in Computer Science (LNCS), vol. 3823, Springer, pp. 945–954, December 2005.

[CMR98] Ran Canetti, Daniele Micciancio, and Omer Reingold, *Perfectly one-way probabilistic hash functions (preliminary version)*, STOC '98: Proceedings of the thirtieth annual ACM symposium on Theory of computing, ACM Press, pp. 131–140, 1998.

[CNIPA07] CNIPA Centro Nazionale per l'Informica nella Pubblica Amministrazione, *Linee Guida per l'impiego dei sistemi RFId nella Pubblica Amministrazione*, http://www.cnipa.gov.it/site/_files/cnipa_quad_30.pdf, February 2007.

[Cra99] Ronald Cramer, *Introduction to Secure Computation*, Lectures on Data Security, Modern Cryptology in Theory and Practice, Summer School, Aarhus, Denmark, July 1998, Springer, pp. 16–62, 1999.

[CZ06] *Kombi mit RFID birgt Riesenpotential*, Computer-Zeitung (2006), no. 20.

[DA99] Tim Dierks and Christopher Allen, *RFC 2246: The TLS Protocol Version 1*, IETF RFC 2246, January 1999.

[DDM03] George Danezis, Roger Dingledine, and Nick Mathewson, *Mixminion: Design of a Type III Anonymous Remailer Protocol*, SP '03: Proceedings of the 2003 IEEE Symposium on Security and Privacy, IEEE Computer Society, pp. 2–15, May 2003.

[DMS04] Roger Dingledine, Nick Mathewson, and Paul Syverson, *Tor: The Second-Generation Onion Router*, Proceedings of the 13th USENIX Security Symposium, August 2004.

[DOD94] DOD Joint Staff, *Joint Pub 1-02: Department of Defense Dictionary of Military and Associated Terms*, 1994.

[Due06a] Gunter Dueck, *Lean Brain Management*, Informatik Spektrum vol. 29 (2006), no. 3, pp. 300–305.

[Due06b] Gunter Dueck, *Panopticon*, Informatik Spektrum vol. 29 (2006), no. 6, pp. 442–446.

[EPC05a] *The Application Level Events (ALE) Specification, Version 1.0*, EPCglobal Inc., 2005.

[EPC05b] *Class 1 Generation 2 UHF Air Interface Protocol Standard Version 1.0.9: "Gen 2"*, EPCglobal Inc., 2005.

[EPC05c] *EPC Generation 1 Tag Data Standards Version 1.1 Rev.1.27*, EPCglobal Inc., 2005.

[EPC05d] *Object Naming Service (ONS) Standard, Version 1.0*, EPCglobal Inc., 2005.

[EPC06a] *EPCglobal Certificate Profile Standard*, EPCglobal Inc., 2006.

[EPC06b] *Reader Protocol Standard, Version 1.1*, EPCglobal Inc., 2006.

[EPIC04] EPIC and Privacy International, *Privacy & Human Rights 2004: An International Survey of Privacy Laws and Developments*, Powell's Books, 2004.

[ES02] Daniel W. Engels and Sanjay E. Sarma, *The reader collision problem*, Proceedings of the IEEE International Conference on System, Man and Cybernetics, vol. 3, October 2002.

[Fel03] Martin Feldhofer, *A Proposal for Authentication Protocol in a Security Layer for RFID Smart Tags*, 2003.

[FEMA03] FEMA: Federal Emergency Management Agency, *Risk Management Series, Reference Manual: Chapter 1 – Asset Value, Threat/Hazard, Vulnerability, and Risk*, December 2003.

[Fin03] Klaus Finkenzeller, *RFID Handbook: Fundamentals and Applications in Contactless Smart Cards and Identification*, John Wiley & Sons, Inc., New York, NY, USA, 2003.

[FM05] Elgar Fleisch and Friedemann Mattern (eds.), *Das Internet der Dinge*, Springer, Berlin, Heidelberg, 2005.

[FR03] Kenneth P. Fishkin and Sumit Roy, *Enhancing RFID privacy via antenna energy analysis*, Technical Memo IRS-TR-03-012, presented at MIT RFID Privacy Workshop, Intel Research, November 2003.

[Gam85] Taher El Gamal, *A public key cryptosystem and a signature scheme based on discrete logarithms*, IEEE Transactions on Information Theory vol. 31 (1985), pp. 469–472.

[Gar02] Simson L. Garfinkel, *An RFID Bill of Rights*, Technology Review, October 2002.

[GCvDD02] Blaise Gassend, Dwaine Clarke, Marten van Dijk, and Srinivas Devadas, *Silicon physical random functions*, CCS '02: Proceedings of the 9th ACM conference on Computer and communications security, ACM Press, pp. 148–160, 2002.

[Gil06] Daniel Gilbert, *If only gay sex caused global warming*, Los Angeles Times, July 2006.

[GILC98] GILC, *Privacy International: Global Internet Liberty Campaign: Privacy and Human Rights, An International Survey of Privacy Laws and Practice*, http://www.gilc.org/privacy/survey/, September 1998.

[GJJS04] Philippe Golle, Markus Jakobsson, Ari Juels, and Paul Syverson, *Universal Re-encryption for Mixnets*, Topics in Cryptology CT-RSA 2004, Lecture Notes in Computer Science (LNCS), Springer, pp. 163–178, 2004.

[GJP05] Simson L. Garfinkel, Ari Juels, and Ravi Pappu, *RFID Privacy: An Overview of Problems and Proposed Solutions*, IEEE Security and Privacy vol. 3 (2005), no. 3, pp. 34–43.

[GlB07] GS1, logicaCMG, and BRIDGE, *European passive RFID Market Sizing 2007-2022*, February 2007.

[GLG+02] Tim Grembowski, Roar Lien, Kris Gaj, Nghi Nguyen, Peter Bellows, Jaraslav Flidr, Tom Lehman, and Brian Schott, *Comparative analysis of the hardware implementations of hash functions SHA-1 and SHA*, Proceedings of the 5th International Information Security Conference, 2002.

[Gol04] Harry Goldstein, *We like to watch*, IEEE Spectrum (2004), pp. 27–30.

[GRS99] David Goldschlag, Michael G. Reed, and Peter F. Syverson, *Onion routing for anonymous and private Internet connections*, Communications of the ACM (USA) vol. 42 (1999), no. 2, pp. 39–41.

[Grz02] Torsten Grzebiela, *Insurability of Electronic Commerce Risks*, Proceedings of the 35th Hawaii Internation Conference on System Sciences, IEEE Computer Society, 2002.

[GS05] Oliver Günther and Sarah Spiekermann, *RFID and the perception of control: the consumer's view*, Communications of the ACM vol. 48 (2005), no. 9, pp. 73–76.

[Han05] Gerhard Hancke, *A Practical Relay Attack on ISO 14443 Proximity Cards*, Manuscript, February 2005.

[HGM06a] Dirk Henrici, Joachim Götze, and Paul Müller, *Hash-based Pseudonymity for Ubiquitous Devices*, Poster proceedings of the

International Conference on Computational Science and its Applications, UASS'06 Ubiquitous Application & Security Service, May 2006.

[HGM06b] Dirk Henrici, Joachim Götze, and Paul Müller, *A Hash-based Pseudonymization Infrastructure for RFID Systems*, Proceedings of the IEEE Security, Privacy and Trust in Pervasive and Ubiquitous Computing International Workshop (SecPerU 2006) at IEEE International Conference on Pervasive Services (ICPS 2006), IEEE Computer Society, June 2006.

[HHS73] *Report of the Secretary's Advisory Committee on Automated Personal Data Systems: Records, Computers and the Rights of Citizens*, United States Department of Health and Human Services, July 1973.

[HJSW06] John Halamka, Ari Juels, Adam Stubblefield, and Jonathan Westhues, *The Security Implications of VeriChip™ Cloning*, Journal of the American Medical Informatics Association (JAMIA) vol. 13 (2006), no. 6, pp. 601–607.

[HM04a] Dirk Henrici and Paul Müller, *Hash-Based Enhancement of Location Privacy for Radio-Frequency Identification Devices Using Varying Identifiers*, PerCom 2004, International Workshop on Pervasive Computing and Communication Security, PerSec 2004 (Ravi Sandhu and Roshan Thomas, eds.), IEEE Computer Society, pp. 149–153, March 2004.

[HM04b] Dirk Henrici and Paul Müller, *Tackling Security and Privacy Issues in Radio Frequency Identification Devices*, Pervasive Computing, Second International Conference, PERVASIVE 2004, Vienna, Austria, April 2004, Proceedings (Alois Ferscha and Friedemann Mattern, eds.), Lecture Notes in Computer Science (LNCS), vol. 3001, Springer, pp. 219–224, 2004.

[HM05] Dirk Henrici and Jochen Müller, *Data Security in Service-Oriented Architectures*, Proceedings of the 19th "DFN Arbeitstagung", GI-Edition Lecture Notes in Informatics (LNI), Gesellschaft für Informatik, May 2005.

[HM08] Dirk Henrici and Paul Müller, *Providing Security and Privacy in RFID Systems Using Triggered Hash Chains*, Sixth Annual IEEE International Conference on Pervasive Computing and Communications, PerCom 2008, IEEE Computer Society, 2008.

[HMM04] Dirk Henrici, Jochen Müller, and Paul Müller, *Sicherheit und Privatsphäre in RFID-Systemen*, DFN-Arbeitstagung über Kommunikationsnetze, Lecture Notes in Informatics (LNI), vol. 55, Gesellschaft für Informatik, pp. 45–60, 2004.

[HRC66] Human Rights Committee (HRC), *International Covenant on Civil and Political Rights*, http://www.ohchr.org/english/law/ccpr.htm, December 1966.

[ICC04] International Chamber of Commerce, *The fight against piracy and counterfeiting of intellectual property*, http://www.iccwbo.org/home/intellectual_property/fight_against_piracy.pdf, June 2004.

[IEEE03] *IEEE 802.15.4-2003: Wireless Medium Access Control (MAC) and Physical Layer (PHY) Specifications for Low-Rate Wireless Personal Area Networks (LR-WPANs)*, Institute of Electrical and Electronics Engineers, 2003.

[IKY02] Sozo Inoue, Shin'ichi Konomi, and Hiroto Yasuura, *Privacy in the Digitally Named World with RFID Tags*, Proceedings of the UBICOMP 2002

Workshop on Socially-informed Design of Privacy-enhancing Solutions in Ubiquitous Computing, September 2002.

[ISO01a] *ISO/IEC 14443: Identification cards – Contactless integrated circuit(s) cards – Proximity cards*, International Organization for Standardization, 2001.

[ISO01b] *ISO/IEC 15693: Identification cards – Contactless integrated circuit(s) cards – Vicinity cards*, International Organization for Standardization, 2001.

[ISO03] *ISO/IEC 14223: Radiofrequency identification of animals – Advanced transponders*, International Organization for Standardization, 2003.

[ISO04] *ISO/IEC 18000: Information technology – Radio frequency identification for item management*, International Organization for Standardization, 2004.

[ISO05] *ISO/IEC 17799: Information technology – Security techniques – Code of practice for information security management*, International Organization for Standardization, 2005.

[ISO94] *ISO 7498-1, ITU-Rec. X.200: Information technology – Open Systems Interconnection – Basic Reference Model: The Basic Model*, International Organization for Standardization, 1994.

[JA03] Anne R. Jacobs and Gregory D. Abowd, *A Framework for Comparing Perspectives on Privacy and Pervasive Technologies*, IEEE Pervasive Computing vol. 02 (2003), no. 4, pp. 78–84.

[JHL02] Xiaodong Jiang, Jason I. Hong, and James A. Landay, *Approximate Information Flows: Socially-Based Modeling of Privacy in Ubiquitous Computing*, UbiComp 2002: Proceedings of the 4th international conference on Ubiquitous Computing, Springer, pp. 176–193, 2002.

[JP03] Ari Juels and Ravikanth Pappu, *Squealing Euros: Privacy Protection in RFID-Enabled Banknotes*, Financial Cryptography – FC'03 (Rebecca N. Wright, ed.), Lecture Notes in Computer Science (LNCS), vol. 2742, IFCA, Springer, pp. 103–121, January 2003.

[JRS03] Ari Juels, Ronald L. Rivest, and Michael Szydlo, *The Blocker Tag: Selective Blocking of RFID Tags for Consumer Privacy*, Conference on Computer and Communications Security – ACM CCS (Vijay Atluri, ed.), Association for Computing Machinery (ACM), ACM Press, pp. 103–111, October 2003.

[Jue04] Ari Juels, *Minimalist Cryptography for Low-Cost RFID Tags*, International Conference on Security in Communication Networks – SCN 2004 (Carlo Blundo and Stelvio Cimato, eds.), Lecture Notes in Computer Science (LNCS), vol. 3352, Springer, pp. 149–164, September 2004.

[Jue05] Ari Juels, *Strengthening EPC Tags Against Cloning*, Manuscript, March 2005.

[Jue06] Ari Juels, *RFID Security and Privacy: A Research Survey*, IEEE Journal on Selected Areas in Communications (2006), no. 2, pp. 381–394.

[JY06] Injoo Jang and Hyeong Seon Yoo, *Pseudorandom Number Generator Using Optimal Normal Basis.*, Proceedings Part III of ICCSA 2006, International Conference on Computational Science and its Applications, Lecture Notes in Computer Science (LNCS), Springer, pp. 206–212, 2006.

[KBC97] Hugo Krawczyk, Mihir Bellare, and Ran Canetti, *RFC 2104: HMAC: Keyed-Hashing for Message Authentication*, Internet Engineering Task Force: RFC 1034, February 1997.

[KC04] Jean Kumagai and Steven Cherry, *Sensors & Sensibility*, IEEE Spectrum (2004), pp. 18–24.

[KEB+07] Tom Karygiannis, Bernard Eydt, Greg Barber, Lynn Bunn, and Ted Phillips, *Guidelines for Securing Radio Frequency Identification (RFID) Systems – Recommendations of the National Institute of Standards and Technology*, http://csrc.nist.gov/publications/nistpubs/800-98/SP800-98_RFID-2007.pdf, April 2007.

[KH06] Khoongming Khoo and Swee-Huay Heng, *New Constructions of Universal Hash Functions Based on Function Sums*, Proceedings Part III of ICCSA 2006, International Conference on Computational Science and its Applications, Lecture Notes in Computer Science (LNCS), Springer, pp. 416–425, 2006.

[Kie06] Daniel Kiefer, *Project thesis (written in German): Entwurf eines Verfahrens zur Eigentums- und Besitzübergabe in RFID-Systemen (designing a procedure for the transfer of possession and ownership in RFID systems)*, Technical report, Workgroup for Integrated Communication Systems, University of Kaiserslautern, 2006.

[KMV00] Neal Koblitz, Alfred J. Menezes, and Scott Vanstone, *The State of Elliptic Curve Cryptography*, Des. Codes Cryptography vol. 19 (2000), no. 2-3, pp. 173–193.

[KS05] Stephen T. Kent and Karen Seo, *Security Architecture for the Internet Protocol*, RFC 4301, December 2005.

[KW05] Ziv Kfir and Avishai Wool, *Picking virtual pockets using relay attacks on contactless smartcard systems*, Conference on Security and Privacy for Emerging Areas in Communication Networks – SecureComm 2005, IEEE, September 2005.

[LAK06] Sangshin Lee, Tomoyuki Asano, and Kwangjo Kim, *RFID Mutual Authentication Scheme based on Synchronized Secret Information*, Symposium on Cryptography and Information Security, January 2006.

[Lam81] Leslie Lamport, *Password authentication with insecure communication*, Communications of the ACM vol. 24 (1981), no. 11, pp. 770–772.

[Lan01] Marc Langheinrich, *Privacy by Design - Principles of Privacy-Aware Ubiquitous Systems*, Proceedings of UbiComp 2001, pp. 273–291, September 2001.

[Lan02] Marc Langheinrich, *Privacy Invasions in Ubiquitous Computing*, Proceedings of UbiComp 2002, Lecture Notes in Computer Science (LNCS), Springer, September 2002.

[Lan05] Marc Langheinrich, *Personal Privacy in Ubiquitous Computing – Tools and System Support*, Ph.D. thesis, ETH Zurich, Zurich, Switzerland, May 2005.

[LDM02] Scott Lederer, Anind K. Dey, and Jennifer Mankoff, *Everyday Privacy in Ubiquitous Computing Environments*, Privacy in Ubicomp 2002, Workshop on Socially-informed Design of Privacy-enhancing Solutions in Ubiquitous Computing (2002).

[LHLL05] Su-Mi Lee, Young Ju Hwang, Dong Hoon Lee, and Jong In Lim, *Efficient Authentication for Low-Cost RFID Systems*, International Conference on Computational Science and its Applications - ICCSA 2005,

Proceedings, Part I (Osvaldo Gervasi, Marina Gavrilova, Vipin Kumar, Antonio Laganaà, Heow Pueh Lee, Youngsong Mun, David Taniar, and Chih Jeng Kenneth Tan, eds.), Lecture Notes in Computer Science (LNCS), vol. 3480, Springer, pp. 619–627, May 2005.

[Lin03] Mats G. Lindquist, *RFID in libraries - introduction to the issues*, World Library and Information Congress: 69th IFLA General Conference and Council, http://www.ifla.org/IV/ifla69/papers/161e-Lindquist.pdf, August 2003.

[LLG⁺04] Jae W. Lee, Daihyun Lim, Blaise Gassend, G. Edward Suh, Marten van Dijk, and Srini Devadas, *A Technique to Build a Secret Key in Integrated Circuits for Identification and Authentication Applications*, Proceedings of the IEEE VLSI Circuits Symposium, June 2004.

[LSMF06] Mikko Lehtonen, Thorsten Staake, Florian Michahelles, and Elgar Fleisch, *From Identification to Authentication - A Review of RFID Product Authentication Techniques*, Printed handout of Workshop on RFID Security – RFIDSec'06, July 2006.

[Mar01] Gary T. Marx, *Murky conceptual waters: The public and the private*, Ethics and Information Technology vol. 3 (2001), no. 3, pp. 157–169.

[Mat01] Friedemann Mattern, *Pervasive/Ubiquitous Computing*, Informatik Spektrum vol. 24 (2001), no. 3, pp. 145–147.

[Mat04] Friedemann Mattern, *Allgegenwärtige Informationstechnik – Soziale Folgen und Konsequenzen für die Menschenrechte*, Menschenrechte und Terrorismus (P. Kirchschläger, Th. Kirchschläger, A. Belliger, and D. Krieger, eds.), Stämpfli Verlag, Bern, Switzerland, 2004, pp. 315–335.

[McF05] Paul McFedries, *Bluetooth Cavities*, IEEE Spectrum (2005), p. 72.

[Mer06] Peter Mertens, *Das Ungleichgewicht im Datenschutz*, Informatik Spektrum vol. 29 (2006), no. 6, pp. 416–423.

[MFD03] Ginger Myles, Adrian Friday, and Nigel Davies, *Preserving Privacy in Environments with Location-Based Applications*, IEEE Pervasive Computing vol. 2 (2003), no. 1, pp. 56–64.

[Mil86] Victor S. Miller, *Use of elliptic curves in cryptography*, Advances in Cryptology Proceedings– CRYPTO'85, Lecture Notes in Computer Science (LNCS), Springer, pp. 417–426, 1986.

[MK98] David Mazières and M. Frans Kaashoek, *The Design, Implementation and Operation of an Email Pseudonym Server*, Proceedings of the 5th ACM Conference on Computer and Communications Security, Association for Computing Machinery (ACM), 1998.

[Moc87] Paul Mockapetris, *RFC 1034: Domain names - concepts and facilities*, Internet Engineering Task Force: RFC 1034, November 1987.

[Moo65] Gordon E. Moore, *Cramming more components onto integrated circuits*, Electronics Magazine; McGraw-Hill vol. 38 (1965), no. 8, pp. 114–117.

[MSW05] David Molnar, Andrea Soppera, and David Wagner, *A Scalable, Delegatable Pseudonym Protocol Enabling Ownership Transfer of RFID Tags*, Selected Areas in Cryptography – SAC 2005 (Bart Preneel and Stafford Tavares, eds.), Lecture Notes in Computer Science (LNCS), vol. 3897, Springer, pp. 276–290, August 2005.

[MvOV96] Alfred J. Menezes, Paul C. van Oorschot, and Scott A. Vanstone, *Handbook of Applied Cryptography*, CRC Press, http://www.cacr.math.uwaterloo.ca/hac/, October 1996.

[NIST02] National Institute of Standards and Technology (NIST), *Secure hash standard*, U.S. Department of Commerce, Washington, August 2002, http://csrc.nist.gov/publications/fips/, Federal Information Processing Standard 180-2.

[Orw49] George Orwell, *Nineteen Eighty-Four*, Secker and Warburg, 1949.

[OSK03] Miyako Ohkubo, Koutarou Suzuki, and Shingo Kinoshita, *Cryptographic Approach to "Privacy-Friendly" Tags*, RFID Privacy Workshop, November 2003.

[PLHCETR06] Pedro Peris-Lopez, Julio Cesar Hernandez-Castro, Juan Estevez-Tapiador, and Arturo Ribagorda, *RFID Systems: A Survey on Security Threats and Proposed Solutions*, 11th IFIP International Conference on Personal Wireless Communications – PWC'06, Lecture Notes in Computer Science (LNCS), vol. 4217, Springer, pp. 159–170, September 2006.

[PRB98a] Bart Preneel, Vincent Rijmen, and Antoon Bosselaers, *Design principles and performance of conventional cryptographic algorithms*, Dr. Dobb's Journal vol. 23 (1998), no. 12, pp. 126–131.

[PRB98b] Bart Preneel, Vincent Rijmen, and Antoon Bosselaers, *Recent developments in the design of conventional cryptographic algorithms*, State of the Art and Evolution of Computer Security and Industrial Cryptography, 1997 (Bart Preneel and Vincent Rijmen, eds.), Lecture Notes in Computer Science (LNCS), vol. 1528, Springer, pp. 106–131, 1998.

[REC04] Damith Ranasinghe, Daniel Engels, and Peter Cole, *Security and Privacy: Modest Proposals for Low-Cost RFID Systems*, Auto-ID Labs Research Workshop, September 2004.

[Rio05] Michael Riordan, *How Europe Missed The Transistor*, IEEE Spectrum (2005), pp. 46–51.

[Riv01] Ronal E. Rivest, *Whither Information Security?*, MIT Laboratory for Computer Science, http://wean1.ulib.org/Lectures/Distinguished%20Lectures/2001/03.0%20Ronald%20L%20Rivest/6SLIDES/security.ppt, 2001.

[Riv92] Ronald L. Rivest, *The MD5 Message-Digest Algorithm*, IETF RFC 1321, April 1992.

[Roß01] Alexander Roßnagel, *Allianz von Medienrecht und Informationstechnik?*, Nomos Verlagsgesellschaft, October 2001.

[Roß02] Alexander Roßnagel, *Freiheit im Cyberspace*, Informatik Spektrum vol. 25 (2002), no. 1, pp. 33–38.

[Roß05] Alexander Roßnagel, *Verantwortung für Datenschutz*, Informatik Spektrum vol. 28 (2005), no. 6, pp. 462–473.

[Rya67] William G. Ryan, *Privacy and freedom*, Elsevier Business Horizons vol. 10 (1967), no. 4, p. 106.

[SA99] Frank Stajano and Ross Anderson, *The Resurrecting Duckling: Security Issues for Ad-hoc Wireless Networks*, Security Protocols, 7th International Workshop Proceedings (Bruce Christianson, Bruno Crispo, and Michael Roe, eds.), Lecture Notes in Computer Science (LNCS), Springer, pp. 172–194, 1999.

[Sat01] Mahadev Satyanarayanan, *Pervasive Computing: Vision and Challenges*, IEEE Personal Communications, pp. 10–17, August 2001.

[Sat03] Mahadev Satyanarayanan, *Privacy: The Achilles Heel of Pervasive Computing?*, IEEE Pervasive Computing vol. 02 (2003), no. 1, pp. 2–3.

[Say94] Mahmoud Sayrafiezadeh, *The Birthday Problem Revisited*, Mathematics Magazine vol. 67 (1994), pp. 220–223.

[Sch01] Tom A. Scharfeld, *An Analysis of the Fundamental Constraints on Low Cost Passive Radio-Frequency Identification System Design*, Master's thesis, Massachusetts Institute of Technology, August 2001.

[Sch03] Bruce Schneier, *Beyond Fear: Thinking Sensibly about Security in an Uncertain World*, Springer, April 2003.

[Sch05] Bruce Schneier, *Mitigating Identity Theft*, Crypto-Gram, April 2005.

[Sch06a] Bruce Schneier, *Casual Conversation, R.I.P.*, Forbes.com, October 2006.

[Sch06b] Bruce Schneier, *The Death of Ephemeral Conversation*, Crypto-Gram, November 2006.

[Sch06c] Bruce Schneier, *Facebook and Data Control*, Crypto-Gram, October 2006.

[Sch06d] Bruce Schneier, *Perceived Risk vs. Actual Risk*, Crypto-Gram, October 2006.

[Sch06e] Bruce Schneier, *Renew Your Passport Now!*, Crypto-Gram, October 2006.

[Sha49] Claude E. Shannon, *Communication Theory of Secrecy Systems*, Bell System Technical Journal vol. 28 (1949), no. 4, pp. 656–715.

[Sha83] Adi Shamir, *On the generation of cryptographically strong pseudo-random sequences*, ACM Transactions on Computer Systems (1983), no. 1, pp. 38–44.

[Smi00] Robert Ellis Smith, *Ben Franklin's Web Site: Privacy and Curiosity from Plymouth Rock to the Internet*, Sheridan Books, Providence, RI, 2000.

[Sol06] Daniel J. Solove, *A Taxonomy of Privacy*, University of Pennsylvania Law Review vol. 154 (2006), no. 3, p. pp. 477ff.

[Ste94] Richard W. Stevens, *TCP/IP Illustrated Volume 1, The Protocols*, Addison-Wesley professional computing series, Addison-Wesley, Cambridge, 1994.

[STF05] Thorsten Staake, Frédéric Thiesse, and Elgar Fleisch, *Extending the EPC Network – The Potential of RFID in Anti-Counterfeiting*, Symposium on Applied Computing – SAC (Hisham Haddad, Lorie Liebrock, Andrea Omicini, and Roger Wainwright, eds.), Association for Computing Machinery, ACM Press, pp. 1607–1612, March 2005.

[Sto03] Adam Stone, *The Dark Side of Pervasive Computing*, IEEE Pervasive Computing vol. 2 (2003), no. 1, pp. 4–8.

[Sur05] Ursula Sury, *Allgegenwärtige Informationstechnik als Herausforderung für die Menschenrechte*, Informatik Spektrum vol. 28 (2005), no. 3, pp. 247–249.

[Sur06] Ursula Sury, *DRM*, Informatik Spektrum vol. 29 (2006), no. 1, pp. 66–68.

[SWE02] Sanjay E. Sarma, Stephen A. Weis, and Daniel W. Engels, *RFID Systems and Security and Privacy Implications*, Cryptographic Hardware and Embedded Systems – CHES 2002 (Burton Kaliski, Çetin Kaya Koç, and Christof Paar, eds.), Lecture Notes in Computer Science (LNCS), vol. 2523, Springer, pp. 454–469, August 2002.

[SWE03] Sanjay E. Sarma, Stephen A. Weis, and Daniel W. Engels, *Radio-Frequency Identification: Security Risks and Challenges*, Cryptobytes, RSA Laboratories vol. 6 (2003), no. 1, pp. 2–9.

[TB06] Pim Tuyls and Lejla Batina, *RFID-Tags for Anti-Counterfeiting*, Topics in Cryptology - CT-RSA 2006, The Cryptographers' Track at the RSA Conference 2006 (David Pointcheval, ed.), Lecture Notes in Computer Science (LNCS), vol. 3860, Springer, pp. 115–131, February 2006.

[UN48] United Nations, *Universal Declaration of Human Rights*, December 1948, adopted and proclaimed by General Assembly resolution, 217 A(III).

[Uro03] Melvin Urofsky, *Rights of the People: Individual Freedom and the Bill of Rights*, http://usinfo.state.gov/products/pubs/rightsof/privacy.htm, December 2003.

[Van04] Scott Vanstone, *Elliptic Curve Crpytography: Die nächste Generation der drahtlosen Sicherheit*, Markt & Technik (2004), no. 17, pp. 20–22.

[Wan06] Roy Want, *An Introduction to RFID Technology*, IEEE Pervasive Computing vol. 5 (2006), no. 1, p. 25.

[WB90] Samuel D. Warren and Louis D. Brandeis, *The Right to Privacy*, Harvard Law Review, vol. IV, Harvard Law Review, no. 5, December 1890.

[Web05] *ePaper RFID Tag*, http://ubiks.net/local/blog/jmt/archives3/004311.html, September 2005.

[Web06] *Website of I2P*, http://www.i2p.net, 2006.

[Wei00] Steve H. Weingart, *Physical Security Devices for Computer Subsystems: A Survey of Attacks and Defences*, CHES 2000: Proceedings of the Second International Workshop on Cryptographic Hardware and Embedded Systems, Lecture Notes in Computer Science (LNCS), vol. 1965, Springer, pp. 302–317, 2000.

[Wei03] Stephen A. Weis, *Security and Privacy in Radio-Frequency Identification Devices*, Master's thesis, Massachusetts Institute of Technology, Cambridge, MA 02139, May 2003.

[Wei05] Paul Weibert, *Project thesis (written in German): Entwicklung einer dezentralen Pseudonymisierungsinfrastruktur (development of a decentralized pseudonymization infrastructure)*, Project thesis (written in german), Workgroup for Integrated Communication Systems, University of Kaiserslautern, 2005.

[Wei91] Mark Weiser, *The Computer for the 21st Century*, Scientific American vol. 265 (1991), no. 3, pp. 94–104.

[Wei93] Mark Weiser, *Some Computer Science Issues in Ubiquitous Computing.*, Communications of the ACM vol. 36 (1993), no. 7, pp. 74–84.

[Whi04] James Q. Whitman, *The Two Western Cultures of Privacy: Dignity versus Liberty*, Yale Law Journal vol. 113 (2004), pp. 1151–1221.

[WHO06] World Health Organization, *Counterfeit medicines, Fact sheet No. 275*, http://www.who.int/mediacentre/factsheets/fs275/en/print.html, November 2006.

[WSRE03] Stephen A. Weis, Sanjay E. Sarma, Ronald L. Rivest, and Daniel W. Engels, *Security and Privacy Aspects of Low-Cost Radio Frequency Identification Systems*, International Conference on Security in Pervasive Computing – SPC 2003 (Dieter Hutter, Günter Müller, Werner Stephan, and Markus Ullmann, eds.), Lecture Notes in Computer Science (LNCS), vol. 2802, Springer, pp. 454–469, March 2003.

[WT99] Alma Whitten and Doug ("J. D.") Tygar, *Why Johnny can't encrypt: A usability evaluation of PGP 5.0*, 8th USENIX Security Symposium, 1999.

[WYY05] Xiaoyun Wang, Yiqun Lisa Yin, and Hongbo Yu, *Finding Collisions in the Full SHA-1*, CRYPTO, Lecture Notes in Computer Science (LNCS), vol. 3621, Springer, pp. 17–36, 2005.

Index

Printed in the United States
124891LV00003B/94-117/P

9 783540 790754